Hippocampus - Cytoarchitecture and Diseases

Edited by Xinhua Zhang

Published in London, United Kingdom

IntechOpen

Supporting open minds since 2005

Hippocampus - Cytoarchitecture and Diseases
http://dx.doi.org/10.5772/intechopen.95716
Edited by Xinhua Zhang

Contributors

Douglas D. Burman, Saeed Vafaei-Nezhad, Masood Vafaei-Nezhad, Mehri Shadi, Samira Ezi, Yifan Luo, Federico Stella, Bailu Si, Matteo Toso, Alessandro Treves, Carlos Matias, Fatima M.C. Bastos, Joao P. Vieira, Rosa M. Santos, Luis M. Rosario, Rosa M. Quinta-Ferreira, Maria Emilia Quinta-Ferreira, Ines O. Lopes, Cheril Tapia-Rojas, Débora Buendía, Claudia Jara, Alvaro O Ardiles, Pablo Muñoz, Wen Li, Guohua Jin, Haoming Li, Siwei Xu, Yaya Ji, Tianle Sha, Ruiying Jing, Qiujie Cai, Xinhua Zhang, Lei Zhang, Yongqi Zhu, Kun Jiang, Dandan Chu, Fei Liu

Notice

Statements and opinions expressed in the chapters are these of the individual contributors and not necessarily those of the editors or publisher. No responsibility is accepted for the accuracy of information contained in the published chapters. The publisher assumes no responsibility for any damage or injury to persons or property arising out of the use of any materials, instructions, methods or ideas contained in the book.

First published in London, United Kingdom, 2022 by IntechOpen
IntechOpen is the global imprint of INTECHOPEN LIMITED, registered in England and Wales, registration number: 11086078, 5 Princes Gate Court, London, SW7 2QJ, United Kingdom
Printed in Croatia

British Library Cataloguing-in-Publication Data
A catalogue record for this book is available from the British Library

Additional hard and PDF copies can be obtained from orders@intechopen.com

Hippocampus - Cytoarchitecture and Diseases
Edited by Xinhua Zhang
p. cm.
Print ISBN 978-1-83969-911-5
Online ISBN 978-1-83969-912-2
eBook (PDF) ISBN 978-1-83969-913-9

Meet the editor

Prof. Xinhua Zhang, Ph.D., is head of the Department of Anatomy, Medical College of Nantong University, China. He obtained his Ph.D. from the Institute of Biochemistry and Cell Biology, Chinese Academy of Sciences, Shanghai. Prof. Zhang's research interests include the translational medicine of stem cells and neurodegeneration disorders including Alzheimer's and Parkinson's diseases. He is a member of the council of the Jiangsu Neuroscience Association and Jiangsu Anatomy Society. He has won several awards, including the outstanding member of Jiangsu Province "333 Talent Project" and Jiangsu Province "Blue Project." His work has won the Natural Science Award from the Ministry of Education, the Chinese Medical Science and Technology Award, and the Jiangsu Science and Technological Progress Award, among others.

Contents

Preface

It has been more than 400 years since the hippocampus was first described. As early as 1578, Arantius described the processes at the bottom of the temporal angle as the hippocampus. Based on the ventricular morphology, in 1732 Winslow suggested using the noun "ram horn" for the structure. In the same period, De Garengeot used the term Cornu Ammonis for the hippocampus, referring to the Egyptian god Amun, who has the head of a ram. Research has made the structure and function of the hippocampus clearer and clearer, but there are still many unknown secrets.

The hippocampus is a bicortical structure composed of the Cornu Ammonis and dentate gyrus. They are interlaced with each other, and separated from each other, bounded by the hippocampal sulcus. The hippocampus belongs to the oldest part of the evolution of the brain, namely, the paleocortex. The Cornu Ammonis has three layers: the molecular, pyramidal, and pleomorphic layers. The cortex of the dentate gyrus is also divided into three layers: the molecular, granular, and pleomorphic layers. In the granular cell layer, the marginal area near the hilus is called the subgranular layer, where in the 1960s and 1970s it was found that there are neural stem cells with self-renewal and multi-differentiation potential in adult mammals. The subgranular layer is one of the few areas of the nervous system where postnatal neurogenesis occurs. Studies have shown that the dentate gyrus of primates still retains proliferative precursor cells in adulthood, but few in the aged. Therefore, the hippocampus has the ability of neurogenesis, and the existence of neural stem cells also provides a basis for hippocampal structure and functional plasticity.

There are extensive fiber connections between the hippocampus and other brain regions. The afferent fibers of the hippocampus mainly include perforating fibers from the entorhinal cortex and septal-hippocampal fibers from the septal nucleus. The fimbria fornix is the main efferent pathway of the hippocampus. In addition to terminating the mammillary body, there are fibers ending in the cingulate gyrus, septal nucleus, preoptic area, lateral hypothalamic area, anterior thalamic nucleus, and so on. Because of these connections, the hippocampus is involved in a variety of functions, such as learning, memory, attention, emotion, sensory information processing, and motor function. The hippocampus plays a crucial role in learning and memory, involving all aspects of narrative memory. Information from the neocortex converges to the entorhinal region and then further reaches the hippocampus. Therefore, the newly acquired information is filtered through the hippocampus before reaching the neocortex. The hippocampus can judge whether it is new or recent information, and the identification of old information mainly depends on the neocortex.

The first part of this book consists of four chapters that introduce the cytoarchitecture and functions of the hippocampus. Chapter 1 by Jing et al. describes the learning and memory-related circuits and fiber connections of the hippocampus. Chapter 2 by Bastos et al. states the correlation between fiber autofluorescence and postsynaptic zinc dynamics of pyramidal CA3 neurons. Chapter 3 by Luo et al. focuses on the pattern formation process of entorhinal cortex grid cells. Chapter 4 by Burman demonstrates the influence of the hippocampus across a variety of cognitive domains.

Hippocampal injury can lead to obvious neurological and mental diseases such as Alzheimer's disease (AD), epilepsy, and so on. AD is a latent progressive neurodegenerative disease. Clinically, it is characterized by comprehensive dementia such as memory impairment, aphasia, apraxia, agnosia, and impairment of visuospatial skills. The characteristic pathological changes include atrophy of the hippocampus and cerebral cortex with β-amyloid deposition, neurofibrillary tangles, and loss of neurons. Hippocampal atrophy is associated with early memory impairment, which indicates the possible occurrence of AD. Acetylcholine (ACh) and choline acetyltransferase (ChAT) in the hippocampus and neocortex of AD patients is decreased significantly, which is considered to be one of the causes of memory and cognitive impairment. At present, there is no specific treatment or reversal of disease progression.

The second part of this book focuses on hippocampal-related diseases, including their pathogenesis and treatment. Chapter 5 by Chu and Liu introduces the role of tau protein in physical conditions and the pathological changes related to neurodegenerative diseases, as well as the treatment research based on tau. Chapter 6 by Xu et al. describes the role of TREM2 and microglia in the occurrence and development of AD. Chapter 7 by Wen and Guohua summarizes AD-related circRNAs. Chapter 8 by Jara et al. focuses on the effects of red-light transcranial LED therapy on age-related hippocampal memory. Chapter 9 by Jiang et al. discusses experimental and clinical research using stem cells to treat AD. The stem cells involved include embryonic stem cells, neural stem cells, mesenchymal stem cells, induced pluripotent stem cells, and others. Finally, Chapter 10 by Vafaei-Nezhad et al. talks about the effects of maternal diabetes on multiple brain regions including the hippocampus before and after birth.

I am honored to have had the chance to work on this book with more than thirty authors. In this book, we attempt to provide the basics of hippocampal physiological function, cytoarchitecture, circuits, diseases, and treatment. I'd like to thank the staff IntechOpen, particularly Ms. Maja Bozicevic, who coordinated the publication of this book with great patience and encouragement. We hope this book will be useful for readers interested in hippocampal formation, disease, and treatment.

Xinhua Zhang
Medical College of Nantong University,
Nantong, China

Section 1

Cytoarchetecture and Functions

Chapter 1

Neural Circuits and Some New Factors Involved in Hippocampal Memory

Ruiying Jing, Qiujie Cai, Wen Li and Xinhua Zhang

Abstract

Humans and other primates have memory, and the hippocampus plays a critical role in this process. The neural circuitry is one of the structural foundations for the hippocampus in exerting memory function. To understand the relationship between the hippocampus and memory, we need to understand neural circuits. Past research has identified several classical neural circuits involved in memory. Although there are challenges with the study of hippocampal neural circuits, research on this topic has continued, and some progress has been made. Here, we discuss recent advances in our understanding of hippocampal neural circuit mechanisms and some of the newly discovered factors that affect memory. Substantial progress has been made regarding hippocampal memory circuits and Alzheimer's disease. However, it is unclear whether these novel findings regarding hippocampal memory circuits hold promise for human memory studies. Additional research on this topic is needed.

Keywords: hippocampal, memory, neural circuits, Alzheimer's disease

1. Introduction

Since the description by Scoville and Milner of profound anterograde amnesia in a patient known as H.M. following bilateral temporal lobe resection [1], the hippocampus and surrounding temporal lobe structures have been extensively studied for their role in memory. Subsequently, our understanding of the neurophysiological bases of hippocampal function was greatly enhanced by two breakthroughs: Bliss and Lomo's finding [2] of activity-dependent long-term potentiation (LTP) of synaptic transmission in the hippocampus, and the discovery of hippocampal place cells and neurons that encoded the spatial position of an animal reported by O'Keefe and Dostrovsky [3]. These discoveries stimulated researchers to study the types of memories related to the neural circuits of the hippocampus. Here, we discuss neural circuits and efferent or afferent fibers related to the hippocampus, including the entorhinal cortex to the hippocampus [4], hippocampus to the prefrontal cortex [5], and vDBChATs-dNGIs [6]. We also discuss LIS1, Fos, SynCAM1, BDNF, Smad3, Oxytocin, and DISC1, factors that influence memory insofar as they relate to the development of memory and memory consolidation. With recent technological advances, causal investigations of specific neural circuits relating to the hippocampus and Alzheimer's disease (AD) have helped us to understand the pathogenesis of AD and improve the clinical treatment of AD patients.

2. Cytoarchitecture and functional characteristics in the hippocampus

The hippocampus is an elongated structure with a longitudinal axis extending in a C-shaped fashion, which can be functionally divided into dorsal, intermediate, and ventral parts [7, 8]. Along the transverse axis, it can be further divided into the CA1, CA3, and dentate gyrus (DG). There is a canonical trisynaptic pathway within the hippocampus, involving information proceeding from the entorhinal cortex (EC) to the DG, then to the CA3, and finally to the output node CA1 [9].

The DG has three layers, including the molecular layer, granule cell layer, and polymorphic cell layer [10]. The molecular layer mainly comprises dendrites of the dentate granule cells and the fibers of the perforant path that originate in the entorhinal cortex. The granule cell layer is the principal cell layer, which is largely composed of densely packed granule cells. The granule cell layer encloses a cellular region and forms the third layer of the DG, which is called the polymorphic cell layer. The DG plays a key role in learning, memory, and adult neurogenesis [11]. This region generates new neurons that are integrated into brain circuits [12].

The CA3 area is the largest in the hippocampus and forms the major route of information flow [13]. One of the most prominent features of the CA3 is that there are extensive interconnections among the principal cells via the circulating collateral fiber system [14]. The axon collateral branches of CA3 pyramidal cells form synapses with the apical dendrites of CA3 pyramidal cells in other regions and spread throughout most of the region to form an associative network [15].

The CA1 area, with its widespread projections, is a key output node of the hippocampal memory circuit, which transfers excitatory information out of the hippocampus proper via direct projections to deep layers of the EC or subiculum [16]. The CA1 is composed of densely packed large pyramidal neurons that play an important role in long-term memory and related spatial tasks and behavior [17]. Human CA1 pyramidal neurons exhibit distinctive morphological complexities, which have important computational implications [18]. Many additional functions have been proposed for the CA1, including novelty detection, input comparison, and enrichment of hippocampal output, possibly by redistributing information from the CA3 to a larger number of output neurons [19].

3. The hippocampus and memory

Memory is the ability to use the past to serve the present or future. Without it, we are destined to enter the eternal present. In the twentieth century, Richard Simon introduced the term "engram" to describe the neural matrix used to store and recall memories [20]. Memory is actually a continuous process between nerve cells [21]. Essentially, a population of neurons is activated that undergoes persistent chemical and/or physical changes to become an engram; the neurons reactivate the engram by cues available at the time of the experience [22]. The criteria and experimental strategies in the study discussed by Morris and colleagues to evaluate synaptic plastic memory have become landmarks in evaluating the importance of the existence of engrams [23, 24].

The hippocampus is important for the storage and retrieval of declarative memories, including memories for perception, imagination, and recall of scenes and events [25, 26]. Studies have shown that spatial memory is closely related to the hippocampus. This is because the "place cells" in the hippocampus make the hippocampus necessary for spatial memory activities [27]. If the hippocampus is damaged, humans may not be able to remember where they have been and how to get to where they want to go. For example, AD is a progressive and neurodegenerative disorder of the cortex and hippocampus, characterized by progressive cognitive decline and a prominent loss of hippocampal-dependent memory [28]. Degeneration of basal forebrain cholinergic neurons is a hallmark of AD. Its function depends on the nerve growth factor (NGF), which is transported retrogradely from the synthetic sites in the cortex and hippocampus [29]. Studies have found that patients with Parkinson's disease (PD) also experience a variety of nonmotor symptoms, the most important being cognitive impairment that in many cases can lead to dementia [30]. There is also evidence that the CA1, CA2–3, CA4-DG, and the subiculum are involved in the poor neurocognitive scores of PD memory caused by impairment. Furthermore, because the CA3 is essential for recall, it is expected that atrophy of the CA3 subregion will also affect the episodic memory recollection process in PD patients.

4. Neural circuits and neurite connections involved in hippocampal learning and memory

4.1 The Papez circuit

In the 1930s, Papez et al. discovered that there was a major circuit in the limbic system, called the Papez circuit, involving hippocampus→fornix→papillary body→papillary thalamic tract→prethalamic nucleus→cingulate gyrus→ hippocampus [31, 32] (**Figure 1**). The hippocampus is the central part of this circuit. This circuit acts as the neural basis for emotional expression. It has been shown that axons transmitting emotional impulses originate from the hippocampus and are projected to the thalamus through the papillary body, where physiological emotional effects such as changes in heartbeat, respiration, and body temperature are produced, and nerve fibers are projected to the cingulate gyrus and the frontal lobe of the cerebral cortex after cell replacement to produce a clear emotional experience. Finally, the signal returns to the hippocampus through the projection of the cortex, and then emotional memory will be generated. Later studies have shown that the Papez circuit is also an important brain structure closely related to learning and memory [33, 34]. Therefore, if this circuit is damaged, it will lead to the amnestic syndrome, and different lesions will produce different forms of memory impairment.

4.2 The trisynaptic circuit

The trisynaptic circuit transmits signals between the entorhinal area and the hippocampus structure (**Figure 2**). This circuit first starts in the cortex of the

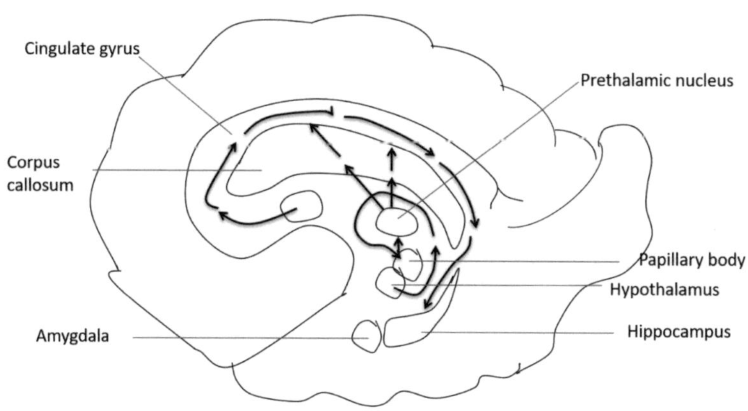

Figure 1.
The Papez circuit.

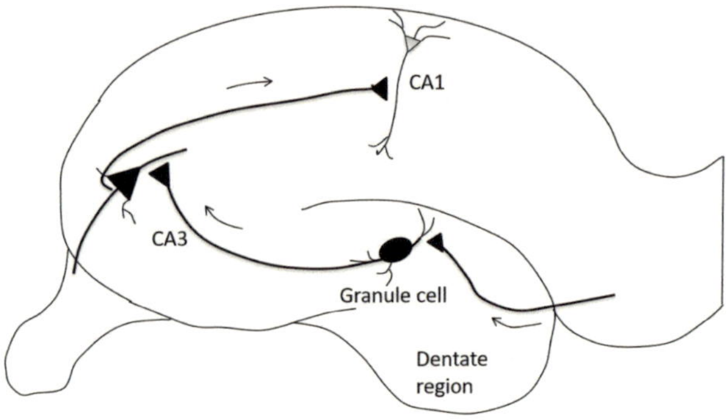

Figure 2.
The trisynaptic circuit.

entorhinal area, where neuronal axons form a perforating circuit and end in the DG granular cell dendrites [35]. The axons of the DG granular cells then form mossy fibers projecting to the hippocampal CA3 area, where they form a second synaptic connection with the dendrites of pyramidal cells. The third synaptic connection occurs between the axons of pyramidal cells in the CA3 area and dendrites of pyramidal cells in the CA1 area, and then the CA1 pyramidal cells transmit the axons to the entorhinal area. The trisynaptic circuit is, therefore, closely related to and forms an important foundation for learning and memory.

4.3 The entorhinal cortex (EC)

The EC is generally perceived as a major input and output structure of hippocampal formation and contributes to cognitive processes and memory formation [36]. The EC is subdivided into two components, the lateral EC (LEC) and medial EC (MEC), according to the differential distribution of EC projections to the DG [37]. In 1893, Santiago Ramon y Cajal [38] described two classical pathways from the EC to the hippocampus (**Figure 3**). One is the long-range pathway: EC layer 2 → DG → CA3 → CA1 area; the other is the short-range or direct pathway: EC layer 3 → CA1 area. However, it is unclear how the hippocampal neurons form specific connection pathways to transmit different information, and how they participate in hippocampal learning functions. Recently, a study [4] found a direct lateral EC-dorsal CA1 (dCA1) circuit that was critically involved in olfactory associative learning. Studies have shown that excitatory pyramidal cells in the hippocampal CA1 region have highly variable molecular, morphological, and electrophysiological

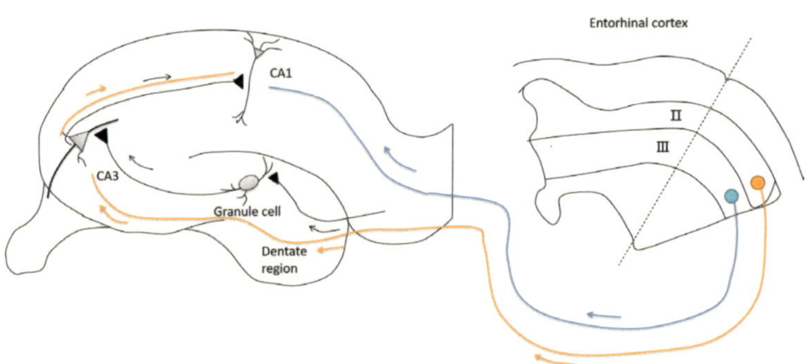

Figure 3.
Pathways from the entorhinal cortex to the hippocampus.

characteristics along the dorso-ventral [39, 40], proximal-distal [40–42], and radial (superficial-deep) axes [39, 40, 43–47]. Subdivisions of deep and superficial pyramidal cells in the CA1 have been recognized for many years, especially along the radial axis. Deep and superficial pyramidal cells are generated at different times and express different genes [39, 40, 43, 48]. *In vivo* recording studies have reported different spiking patterns between these two sublayers; deep cells are more likely to burst and exhibit more spatially tuned firing than superficial cells, and they spike differentially in the hippocampal theta rhythm and during sharp-wave ripple activities [46]. Further study revealed that inhibition of the excitatory synaptic transmission from the LEC to CA1 complex pyramidal cells, or the discharge activity of the CA1 complex pyramidal cells using optogenetic methods, significantly delayed the olfactory association during mice learning [49]. The optogenetic method mentioned above is injecting NpHR or Arch into some mice's hemispheres and then using yellow illumination to identify the mice that include NpHR or Arch. Then researchers can make a comparison between special mice and normal mice. The study also implanted optetrodes into mices' dCA1 to record the olfactory-related firing activities of neurons in the CA1 region of the hippocampus, it was found that the firing of complex pyramidal cells established different preferences for odor cues during learning [49]. These experimental findings revealed that there was a specific neural pathway involved in brain-related learning in the classic circuit from the EC to hippocampus involving cells, synaptic connections, learning behaviors, and neural discharges.

4.4 The prefrontal cortex

The hippocampus and the prefrontal cortex are closely related to advanced cognitive functions of the brain such as learning and memory [50]. Previous studies showed that neural projections from the hippocampus to the prefrontal cortex had the characteristics of a single synapse, were unidirectional, and displayed ipsilateral projections [51]. In brief, the hippocampus-prefrontal lobe projection mainly originates from the subiculum of the ventral hippocampus and the CA1 and ends in the medial, orbitofrontal, and lateral parts of the prefrontal lobe (**Figure 4**). The projection from the prefrontal lobe to the hippocampus is indirectly from the prefrontal lobe to the cingulate gyrus, parahippocampal gyrus, entorhinal cortex, then to the hippocampus [52], which transmits information from prefrontal lobes to important nuclei of the hippocampus [53, 54]. There are also reports that some nerve fibers in the prefrontal lobe project directly to the hippocampus, but the number was low [55, 56].

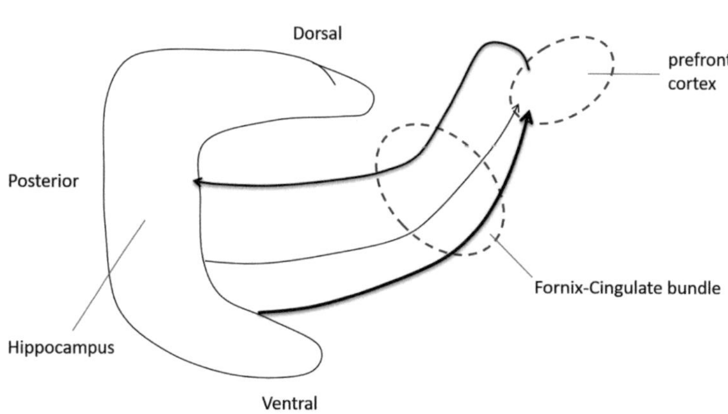

Figure 4.
The pathway from the hippocampus to the prefrontal cortex.

Humans are faced with different environments every day and need to make the right choices through learning and memory in order to find their goals. A recent study [5] found that when an animal runs a specific route and then stops to rest or sleep, place cells [3] are repeatedly released in the same (forward) or opposite (reverse) order as when it ran, and at a faster rate than animals' running. This phenomenon is called memory replay, and this replay has a very important role in the prefrontal cortex circuit which helps the animal remember the path it has taken in the past and makes the right choice among multiple alternative paths.

To understand hippocampus-prefrontal cooperative activities during memory replay and whether memory replay affects animal learning and decision-making, researchers trained rats to learn to find their way in a W-maze [57]. In this task, the rat needs to learn two different rules to get the reward, a simple task and a complex task—in the simple task, the animal only needs to remember the beginning and end positions, then they can successfully find and get the reward; in the complex task, the animal needs to remember the path it has just run and then select the path it has not passed yet among the two available paths to get a reward, this process requires working memory. In this task, most of the memory replay occurred when the animals stayed at the reward site, had just completed a path, or were about to choose the next path [58–61]. The study found that the hippocampus was involved in the processes of both reverse and forward replay. Further studies analyzed whether the content of the hippocampus forward and reverse memory replays changed with learning. They found that the content of forward and reverse replay is different in the different learning stages. In the early stage of learning, reverse replay preferred to the paths that the animal had actually passed before, and thus researchers could accurately predict the animal's past choices from the content of the hippocampal reverse replay. In contrast, forward replay referred to the pathway that the animal will choose next, but this correspondence does not become apparent until later in learning. The dynamic processes of hippocampal reverse and forward replays in the learning process showed that reverse memory replay was very important for animals to remember the past path for learning, and the forward replay was very important for action planning after zoological learning [62, 63]. Besides the W-maze, we can also use the Barnes maze [64] to get the same conclusion. In summary, this study first distinguished the different functions of hippocampal reverse and forward memory replays in spatial learning memory tasks. Reverse memory replay helps to weigh and remember the path to the goal in the past, while forward memory replay is important for planning actions in the future. Moreover, this study for the first time quantitatively described the replay of cooperative memory between the prefrontal cortex and the hippocampus and confirmed its association with animal behavioral choices, to suggest a possible mechanism for the prefrontal cortex to participate in spatial learning.

4.5 vDBChATs-dNGIs

Acetylcholine modifies neuronal excitability, alters presynaptic neurotransmitter release, and coordinates the firing of groups of neurons [65–67]. Recently, researchers used optogenetics, single synaptic tracing, and electrophysiological recording techniques to show that cholinergic neurons in the vertical diagonal band of Broca (vDBChATs) and newly generated immature neurons (NGIs) in the dorsal hippocampus (dNGIs) of adult mice formed a single synaptic connection (vDBChATs-dNGIs synaptic connection) (**Figure 5**); this synaptic transmission was essential for the survival of dNGIs, andthe vDBChATs directly innervate dNGIs. This circuit is mediated by muscarinic cholinergic receptor 1 (M1) on neonatal neurons [68]. In this study, researchers injected a kind of virus that includes mCherry

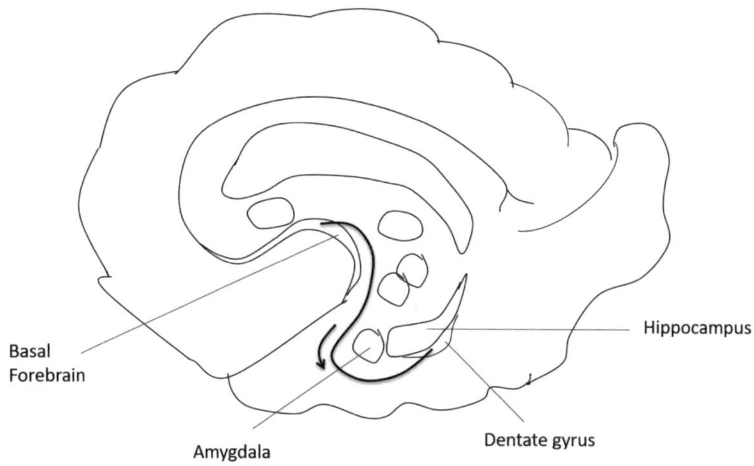

Figure 5.
The vDBChATs-dNGIs circuit.

into the vDB region of some mice. Three days later, in the dDG, mCherry was exclusively expressed in a group of granular cells that were predominately located in the inner one-third of the granule cell layer. Most of these mCherry+ cells expressed doublecortin (DCX, mCherry+DCX+). DCX has been widely established as a marker of immature neurons [69], so mCherry+DCX+ cells are one kind of newly generated NGIs in the dNGIs mentioned above. The study found that using optogenetic technology to enhance vDBChATs-dNGIs synaptic transmission improved spatial learning memory. Furthermore, in the AD transgenic mouse model, the use of optogenetic technology to enhance the synaptic transmission of the neural circuit saved the spatial memory loss of the model mice [70].

5. Factors affecting hippocampal learning and memory

5.1 LIS1

An interesting candidate molecule supporting synaptic integrity is LIS1, which is related to lissencephaly [71, 72]. LIS1 deficits in specific hippocampal neuron populations significantly changed the excitatory synaptic transmission in adult-born Lis1+/− DG projection neurons and dendritic spine density and excitatory synaptic aggregation on hippocampal CA1 projection neurons that lost Lis1 expression from postnatal 20 days [73, 74]. Moreover, the loss of LIS1 after childhood destroys the structure and cell composition of the hippocampus, the connection with other brain regions, and the dependence on the cognition of hippocampal circuits [75, 76].

5.2 Fos

Increasing evidence has shown that sparse neuron groups distributed in many areas of the brain constitute the neural matrix of various behaviors [22, 77]. One sign of these active neuron sets is the instantaneous expression of a group of genes called immediate early genes, one of which encodes the Fos transcription factor, composed of eight members with at least partial functional redundancy (Fos, Fosb, c-Fos, Fosl1, Fosl2, Jun, Junb, and Jund) [78–81]. A long-standing hypothesis is that once activated by a significant stimulation, the neurons expressing Fos will undergo modification, which is helpful to encode specific experience characteristics, so that even if a subset of these neurons are subsequently reactivated, it is enough to trigger memories of the initial experience [82]. Compared with non-Fos-activated neurons, Fos-activated neurons in the hippocampal CA1 region have been shown to stably encode context information [77].

5.3 SynCAM 1

The expression of the synaptic cell adhesion molecule, SynCAM 1, in forebrain neurons, which is also known as a cell adhesion molecule 1 (Cadm1) and Necl-2, is a candidate protein used to evaluate the role of different regions of synaptic tissue proteins [83]. SynCAM 1 belongs to four homophilic and heterophilic membrane protein families of the immunoglobulin superfamily, which are expressed at the peak of synaptic formation and exist until adulthood. This marks the edge of excitatory postsynaptic sites, which is sufficient to induce functional excitatory presynaptic specialization [84]. Studies on knockout and overexpression of the hippocampal CA1 region in mice have shown that SynCAM 1 is necessary to promote excitatory synaptic input of excitatory neurons in vivo [85]. SynCAM 1 also accelerated synapse maturation, which improved the stability of newly formed synapses and in turn increased the likelihood of survival of adult-born neurons [86]. SynCAM 1, therefore, regulates the input of excitatory mossy fibers into the interneurons and major neurons in the hippocampal CA3 region to balance network excitability [87].

5.4 Brain-derived neurotrophic factor (BDNF)

BDNF is one of the most widely distributed and studied neurotrophic factors in mammalian brains. It has a direct impact on memory through various mechanisms. BDNF regulates many different cellular processes involved in the maintenance and development of normal brain function, by binding and activating the TrkB, which is a member of the larger family of Trk receptors [88]. For example, during embryogenesis, BDNF–TrkB signaling promotes the differentiation of cortical progenitor cells and then promotes differentiation of cortical progenitor cells into neurons (i.e., neurogenesis) [89]. The single nucleotide polymorphism of BDNF most likely affects memory through long-term potentiation (LTP), which is important for memory persistence [90]. In the human BDNF gene, a single nucleotide polymorphism leads to an amino acid substitution of valine (Val66Val) to methionine at amino acid residue 66 (Val66Met), which changes the secretion of the mature peptide. This alteration has been related to cognitive deficits among carriers [91]. The effects of BDNF on LTP are also mediated by the TrkB receptor. Especially in the hippocampus, this neurotrophin is thought to act on both pre and postsynaptic compartments, modulating synaptic efficacy, not only by changing the presynaptic transmitter release but also by increasing postsynaptic transmitter sensitivity to induce a long-lasting increase in synaptic plasticity [92, 93].

In the elderly with normal cognition, the presence of BDNF Val66Met is associated with greater hippocampal atrophy and faster cognitive decline [94]. BDNF polymorphism is associated with larger DG volumes within the anterior hippocampus (head) in Met-carriers compared to Val/Val homozygotes. The total hippocampal volume predicted the performance on visuospatial memory tasks in Met-carriers [95]. Although little is known about the process of memory consolidation, it is known that a hippocampal BDNF-positive autoregulatory feedback loop is necessary to mediate memory consolidation via the CCAAT-enhancer-binding protein β (C/EBPβ) [96].

BDNF also mediates the influence of many factors on memory. First, TLQP62, which is a neuropeptide derived from the neurotrophin-inducible VGF (nonacronymic) protein, is capable of inducing increased memory in the mouse hippocampus by promoting neurogenesis and synaptic plasticity through BDNF and its receptor tyrosine receptor kinase B (TrkB) [97, 98]. When TLQP62 promotes BDNF expression, which in turn activates the BDNF/TrkB/CREB (cAMP response element-binding protein) pathway that upregulates VGF expression, there is a VGF-BDNF regulatory loop that appears to regulate neurogenesis [99]. In addition, as is well known, exercise can promote the formation of memory, which is also inseparable from BDNF levels. Lactate, a metabolite released during exercise by muscles, crosses the blood–brain barrier and accumulates in the hippocampus, where it promotes the formation of learning and memory by inducing BDNF expression through silent information regulator 1-dependent induction of the PGC1a/FNDC5 pathway [100]. In addition, the increase of the microglia-dependent proBDNF/BDNF ratio following persistent inflammatory pain leads to cell death of the CA1 and DG neurons. Then, this subsequently causes a cognitive deficit in learning and spatial memory functions [29]. Furthermore, in postmenopausal women, the lower plasma BDNF levels are associated with significantly worse memory performance and changes in the function of the working memory circuit [101].

5.5 Smad3

Smad3 is an intracellular molecule involved in the transforming growth factor-β signaling cascade, which is strongly expressed by granulosa cells of the DG of adult mice [102]. Smad3 deficiency promotes dopaminergic neurodegeneration and α-synuclein aggregation in substantia nigra striatum [103]. Endogenous Smad3 signaling plays important role in neurogenesis and LTP induction of adult DG, which are two forms of hippocampal plasticity related to learning and memory, and which decrease with age and neurological diseases [102].

5.6 Oxytocin

Oxytocin is a brain plasticity regulator of neuronal growth factors, cytoskeleton proteins, and behavioral changes, and is important for short-term hippocampal-dependent memory [104] and regulates neuronal excitability, network oscillatory activity, synaptic plasticity, and society memory [105]. In the SH3 domain and ankyrin repeat-containing, the protein 3 (SHANK3) deficient model related to autism, abnormal neuronal morphology and altered synaptic protein levels are recovered by oxytocin [106]. Early changes of the oxytocin signal may interfere with the maturation of neurons and could have both short-term and long-term pathological consequences [107]. At the molecular level, neurodevelopmental disorders include numerous changes in cytoskeleton rearrangement and neurogenesis, leading to various synaptic diseases [108].

5.7 Disrupted-in-schizophrenia 1 (DISC1)

DISC1 is a strong candidate susceptibility gene for a series of neuropsychiatric diseases [109, 110]. Reports of both DISC1 point mutations (L100P and Q31L) heterozygotes and DISC1 transgenic mice [111, 112] found that the combination of adolescent isolation (from 5 to 8 weeks) and DISC1 L100P mutation damaged the social memory of adults. In addition, adolescent isolation aggravates adult neurogenesis defects in the hippocampus of L100P mice, but has no similar effect on WT mice, and leads to long-term continuous changes in synaptic transmission and plasticity of the hippocampal network of DISC1 L100P mice [113, 114]. There is also a possible sex-dependent effect of DICS1. In the test of significant gene–environment interactions in the amphetamine-induced anxiety in male animals and the amphetamine-induced locomotion in female animals, we surprisingly found that gene–environment interactions improved social memory in not only male but also female animals, but JIA alone disrupted spatial memory and recognition memory only in male animals [115].

6. Expectations

The hippocampus, as an important part of the limbic system involved in learning and memory, has been extensively studied for many years. With increased aging in China, the incidence of AD, a progressive degenerative brain disease, is increasing every year. The main clinical symptoms are memory loss and cognitive impairment. Entropic cortex to the hippocampus, hippocampus to the prefrontal cortex, and vDBChATs-dNGIs with the hippocampus as the central link may play important role in spatial memory and declarative memory. Moreover, the damage of any link in the cycle leads to the loss of recent memory. By studying the hippocampal memory circuit and various influencing factors, we hope to improve spatial memory and declarative memory by intervening in every link of the hippocampal memory circuit. At the same time, we can provide new ideas and methods for the treatment of memory impairment-related diseases such as AD, which are helpful to the recovery and improvement of memory function in the damaged hippocampus. Considering the influence of BDNF and other factors on the memory circuit and the effects of various diseases related to memory impairment, we should also extensively study some influencing factors as intervention targets for Huntington's disease, depression, schizophrenia, bipolar disorder, and other diseases. These studies can provide prevention strategies and treatment methods for memory decline caused by force majeure factors such as sex and age. Furthermore, studies on its influencing factors will open other research avenues.

Author details

Ruiying Jing[1†], Qiujie Cai[1†], Wen Li[1] and Xinhua Zhang[1,2,3*]

1 Department of Human Anatomy, Institute of Neurobiology, Medical School of Nantong University, Nantong, Jiangsu, P.R. China

2 Key Laboratory of Neuroregeneration of Jiangsu and Ministry of Education, Nantong, Jiangsu, P.R. China

3 Co-Innovation Center of Neuroregeneration, Medical School of Nantong University, Nantong, Jiangsu, P.R. China

*Address all correspondence to: zhangxinhua@ntu.edu.cn

† Both authors contributed equally.

References

[1] Scoville WB, Milner B. Loss of recent memory after bilateral hippocampal lesions. Journal of Neurology, Neurosurgery, and Psychiatry. 1957;**20**(1): 11-21

[2] Bliss TV, Lomo T. Long-lasting potentiation of synaptic transmission in the dentate area of the anaesthetized rabbit following stimulation of the perforant path. The Journal of Physiology. 1973;**232**(2):331-356

[3] O'Keefe J, Dostrovsky J. The hippocampus as a spatial map. Preliminary evidence from unit activity in the freely-moving rat. Brain Research. 1971;**34**(1):171-175

[4] Li Y et al. A distinct entorhinal cortex to hippocampal CA1 direct circuit for olfactory associative learning. Nature Neuroscience. 2017;**20**(4):559-570

[5] Shin JD, Tang W, Jadhav SP. Dynamics of awake hippocampal-prefrontal replay for spatial learning and memory-guided decision making. Neuron. 2019;**104**(6):1110-1125.e7

[6] Zhu H et al. Impairments of spatial memory in an Alzheimer's disease model via degeneration of hippocampal cholinergic synapses. Nature Communications. 2017;**8**(1):1676

[7] Santa-Marinha L et al. Phospholipase D1 ablation disrupts mouse longitudinal hippocampal axis organization and functioning. Cell Reports. 2020;**30**(12): 4197-4208.e6

[8] Milior G et al. Electrophysiological properties of CA1 pyramidal neurons along the longitudinal axis of the mouse hippocampus. Scientific Reports. 2016; **6**:38242

[9] Hitti FL, Siegelbaum SA. The hippocampal CA2 region is essential for social memory. Nature. 2014;**508**(7494): 88-92

[10] Amaral DG, Scharfman HE, Lavenex P. The dentate gyrus: Fundamental neuroanatomical organization (dentate gyrus for dummies). Progress in Brain Research. 2007;**163**:3-22

[11] Koyanagi I et al. Memory consolidation during sleep and adult hippocampal neurogenesis. Neural Regeneration Research. 2019;**14**(1):20-23

[12] Kozareva DA, Cryan JF, Nolan YM. Born this way: Hippocampal neurogenesis across the lifespan. Aging Cell. 2019;**18**(5):e13007

[13] Fogwe LA, Reddy V, Mesfin FB. Neuroanatomy, Hippocampus. Treasure Island (FL): StatPearls; 2021

[14] Gilbert PE, Brushfield AM. The role of the CA3 hippocampal subregion in spatial memory: A process oriented behavioral assessment. Progress in Neuro-Psychopharmacology & Biological Psychiatry. 2009;**33**(5):774-781

[15] Scharfman HE, MacLusky NJ. Sex differences in hippocampal area CA3 pyramidal cells. Journal of Neuroscience Research. 2017;**95**(1-2):563-575

[16] Cenquizca LA, Swanson LW. Spatial organization of direct hippocampal field CA1 axonal projections to the rest of the cerebral cortex. Brain Research Reviews. 2007;**56**(1):1-26

[17] Vu T, Gugustea R, Leung LS. Long-term potentiation of the nucleus reuniens and entorhinal cortex to CA1 distal dendritic synapses in mice. Brain Structure & Function. 2020;**225**(6): 1817-1838

[18] Benavides-Piccione R et al. Differential structure of hippocampal CA1 pyramidal neurons in the human and mouse. Cerebral Cortex. 2020;**30**(2):730-752

[19] Soltesz I, Losonczy A. CA1 pyramidal cell diversity enabling parallel information processing in the hippocampus. Nature Neuroscience. 2018;**21**(4):484-493

[20] Josselyn SA, Köhler S, Frankland PW. Heroes of the Engram. The Journal of Neuroscience. 2017; **37**(18):4647-4657

[21] Wang JH. Searching basic units in memory traces: Associative memory cells. F1000Res. 2019;**8**:457

[22] Josselyn SA, Tonegawa S. Memory engrams: Recalling the past and imagining the future. Science. 2020;**367** (6473):p. eaaw4325.

[23] Martin SJ, Grimwood PD, Morris RG. Synaptic plasticity and memory: An evaluation of the hypothesis. Annual Review of Neuroscience. 2000;**23** :649-711

[24] Szu JI, Binder DK. The role of astrocytic aquaporin-4 in synaptic plasticity and learning and memory. Frontiers in Integrative Neuroscience. 2016;**10**:8

[25] Squire LR. Memory systems of the brain: A brief history and current perspective. Neurobiology of Learning and Memory. 2004;**82**(3):171-177

[26] Yang Y, Wang JZ. From structure to behavior in basolateral amygdala-hippocampus circuits. Front Neural Circuits. 2017;**11**:86

[27] Gupta et al. Restrained dendritic growth of adult-born granule cells innervated by transplanted fetal GABAergic interneurons in mice with temporal lobe epilepsy. eNeuro. 2019;**6**(2):p. ENEURO.0110-18.2019.

[28] Gomes BAQ et al. Neuroprotective mechanisms of resveratrol in Alzheimer's disease: Role of SIRT1. Oxidative Medicine and Cellular Longevity. 2018;**2018**:8152373

[29] Fahnestock M, Shekari A. ProNGF and neurodegeneration in Alzheimer's disease. Frontiers in Neuroscience. 2019;**13**:129

[30] Das T, Hwang JJ, Poston KL. Episodic recognition memory and the hippocampus in Parkinson's disease: A review. Cortex. 2019;**113**:191-209

[31] Ferreira TA Jr et al. Postmortem dissections of the Papez circuit and nonmotor targets for functional neurosurgery. World Neurosurgery. 2020;**144**:e866-e875

[32] Weininger J et al. Papez's forgotten tract: 80 years of unreconciled findings concerning the thalamocingulate tract. Frontiers in Neuroanatomy. 2019;**13**:14

[33] LeDoux JE. Emotional memory systems in the brain. Behavioural Brain Research. 1993;**58**(1-2):69-79

[34] Aggleton JP, Brown MW. Episodic memory, amnesia, and the hippocampal-anterior thalamic axis. The Behavioral and Brain Sciences. 1999;**22**(3):425-444 discussion 444-89

[35] Laatsch RH, Cowan WM. Electron microscopic studies of the dentate gyrus of the rat. I. Normal structure with special reference to synaptic organization. The Journal of Comparative Neurology. 1966;**128**(3):359-395

[36] Olajide OJ, Suvanto ME, Chapman CA. Molecular mechanisms of neurodegeneration in the entorhinal cortex that underlie its selective vulnerability during the pathogenesis of Alzheimer's disease. Biology Open. 2021;**10**(1):bio056796

[37] Witter MP et al. Architecture of the entorhinal cortex A review of entorhinal anatomy in rodents with some comparative notes. Frontiers in Systems Neuroscience. 2017;**11**:46

[38] Baratas Díaz LA. Historical meanings of Santiago Ramón y Cajal's

Rétine des vertébrés. Summary of his first scientific. Asclepio. 1994;**46**(1):243-259

[39] Dong HW et al. Genomic-anatomic evidence for distinct functional domains in hippocampal field CA1. Proceedings of the National Academy of Sciences of the United States of America. 2009; **106**(28):11794-11799

[40] Cembrowski MS et al. Spatial gene-expression gradients underlie prominent heterogeneity of CA1 pyramidal neurons. Neuron. 2016;**89**(2): 351-368

[41] Graves AR et al. Hippocampal pyramidal neurons comprise two distinct cell types that are countermodulated by metabotropic receptors. Neuron. 2012;**76**(4):776-789

[42] Henriksen EJ et al. Spatial representation along the proximodistal axis of CA1. Neuron. 2010;**68**(1):127-137

[43] Baimbridge KG et al. Bursting response to current-evoked depolarization in rat CA1 pyramidal neurons is correlated with lucifer yellow dye coupling but not with the presence of calbindin-D28k. Synapse. 1991;7(4): 269-277

[44] Lee SH et al. Parvalbumin-positive basket cells differentiate among hippocampal pyramidal cells. Neuron. 2014;**82**(5):1129-1144

[45] Mizuseki K et al. Hippocampal CA1 pyramidal cells form functionally distinct sublayers. Nature Neuroscience. 2011;**14**(9):1174-1181

[46] Valero M et al. Determinants of different deep and superficial CA1 pyramidal cell dynamics during sharp-wave ripples. Nature Neuroscience. 2015;**18**(9):1281-1290

[47] Danielson NB et al. Sublayer-specific coding dynamics during spatial navigation and learning in hippocampal area CA1. Neuron. 2016;**91**(3):652-665

[48] Slomianka L et al. Hippocampal pyramidal cells: The reemergence of cortical lamination. Brain Structure & Function. 2011;**216**(4):301-317

[49] Igarashi KM et al. Coordination of entorhinal-hippocampal ensemble activity during associative learning. Nature. 2014;**510**(7503):143-147

[50] Preston AR, Eichenbaum H. Interplay of hippocampus and prefrontal cortex in memory. Current Biology. 2013;**23**(17):R764-R773

[51] Laroche S, Davis S, Jay TM. Plasticity at hippocampal to prefrontal cortex synapses: Dual roles in working memory and consolidation. Hippocampus. 2000;**10**(4):438-446

[52] Goldman-Rakic PS, Selemon LD, Schwartz ML. Dual pathways connecting the dorsolateral prefrontal cortex with the hippocampal formation and parahippocampal cortex in the rhesus monkey. Neuroscience. 1984;**12**(3):719-743

[53] Vertes RP. Analysis of projections from the medial prefrontal cortex to the thalamus in the rat, with emphasis on nucleus reuniens. The Journal of Comparative Neurology. 2002;**442**(2): 163-187

[54] Vertes RP et al. Nucleus reuniens of the midline thalamus: Link between the medial prefrontal cortex and the hippocampus. Brain Research Bulletin. 2007;**71**(6):601-609

[55] Hurley KM et al. Efferent projections of the infralimbic cortex of the rat. The Journal of Comparative Neurology. 1991;**308**(2):249-276

[56] Sesack SR et al. Topographical organization of the efferent projections of the medial prefrontal cortex in the

rat: An anterograde tract-tracing study with Phaseolus vulgaris leucoagglutinin. The Journal of Comparative Neurology. 1989;**290**(2):213-242

[57] Fernández-Ruiz A et al. Long-duration hippocampal sharp wave ripples improve memory. Science. 2019;**364**(6445):1082-1086

[58] Ambrose RE, Pfeiffer BE, Foster DJ. Reverse replay of hippocampal place cells is uniquely modulated by changing reward. Neuron. 2016;**91**(5):1124-1136

[59] Diba K, Buzsaki G. Forward and reverse hippocampal place-cell sequences during ripples. Nature Neuroscience. 2007;**10**(10):1241-1242

[60] Foster DJ, Wilson MA. Reverse replay of behavioural sequences in hippocampal place cells during the awake state. Nature. 2006;**440**(7084):680-683

[61] Wu CT et al. Hippocampal awake replay in fear memory retrieval. Nature Neuroscience. 2017;**20**(4):571-580

[62] Wilber AA et al. Laminar organization of encoding and memory reactivation in the parietal cortex. Neuron. 2017;**95**(6):1406-1419.e5

[63] Pfeiffer BE, Foster DJ. Hippocampal place-cell sequences depict future paths to remembered goals. Nature. 2013;**497**(7447):74-79

[64] Pitts MW. Barnes Maze procedure for spatial learning and memory in mice. Bio-Protocol. 2018;**8**(5):p. e2744.

[65] Picciotto MR, Higley MJ, Mineur YS. Acetylcholine as a neuromodulator: Cholinergic signaling shapes nervous system function and behavior. Neuron. 2012;**76**(1):116-129

[66] Gu Z, Yakel JL. Timing-dependent septal cholinergic induction of dynamic hippocampal synaptic plasticity. Neuron. 2011;**71**(1):155-165

[67] Buchanan KA et al. Facilitation of long-term potentiation by muscarinic M(1) receptors is mediated by inhibition of SK channels. Neuron. 2010;**68**(5): 948-963

[68] Hasselmo ME. The role of acetylcholine in learning and memory. Current Opinion in Neurobiology. 2006;**16**(6):710-715

[69] Brown JP et al. Transient expression of doublecortin during adult neurogenesis. The Journal of Comparative Neurology. 2003;**467**(1):1-10

[70] Roy DS et al. Memory retrieval by activating engram cells in mouse models of early Alzheimer's disease. Nature. 2016;**531**(7595):508-512

[71] Reiner O et al. Isolation of a Miller-Dieker lissencephaly gene containing G protein beta-subunit-like repeats. Nature. 1993;**364**(6439):717-721

[72] Lo Nigro C et al. Point mutations and an intragenic deletion in LIS1, the lissencephaly causative gene in isolated lissencephaly sequence and Miller-Dieker syndrome. Human Molecular Genetics. 1997;**6**(2):157-164

[73] Hunt RF et al. LIS1 deficiency promotes dysfunctional synaptic integration of granule cells generated in the developing and adult dentate gyrus. The Journal of Neuroscience. 2012;**32**(37):12862-12875

[74] Sudarov A et al. Lis1 controls dynamics of neuronal filopodia and spines to impact synaptogenesis and social behaviour. EMBO Molecular Medicine. 2013;**5**(4):591-607

[75] Tsien JZ et al. Subregion- and cell type-restricted gene knockout in mouse brain. Cell. 1996;**87**(7):1317-1326

[76] Sudarov A et al. Mature hippocampal neurons require LIS1 for synaptic integrity: Implications for

cognition. Biological Psychiatry. 2018;**83**(6):518-529

[77] Tanaka KZ et al. The hippocampal engram maps experience but not place. Science. 2018;**361**(6400):392-397

[78] Greenberg ME, Ziff EB. Stimulation of 3T3 cells induces transcription of the c-fos proto-oncogene. Nature. 1984;**311**(5985):433-438

[79] Yap EL, Greenberg ME. Activity-regulated transcription: Bridging the gap between neural activity and behavior. Neuron. 2018;**100**(2):330-348

[80] Gallo FT et al. Immediate early genes, memory and psychiatric disorders: Focus on c-Fos, Egr1 and Arc. Frontiers in Behavioral Neuroscience. 2018;**12**:79

[81] Gazon H et al. Hijacking of the AP-1 signaling pathway during development of ATL. Frontiers in Microbiology. 2017;**8**:2686

[82] Yap EL et al. Bidirectional perisomatic inhibitory plasticity of a Fos neuronal network. Nature. 2021;**590**(7844):115-121

[83] Thomas LA, Akins MR, Biederer T. Expression and adhesion profiles of SynCAM molecules indicate distinct neuronal functions. The Journal of Comparative Neurology. 2008;**510**(1): 47-67

[84] Perez de Arce K et al. Topographic mapping of the synaptic cleft into adhesive nanodomains. Neuron. 2015;**88**(6):1165-1172

[85] Robbins EM et al. SynCAM 1 adhesion dynamically regulates synapse number and impacts plasticity and learning. Neuron. 2010;**68**(5):894-906

[86] Doengi M et al. SynCAM 1 improves survival of adult-born neurons by accelerating synapse maturation. Hippocampus. 2016;**26**(3):319-328

[87] Park KA et al. Excitatory synaptic drive and feedforward inhibition in the hippocampal CA3 circuit are regulated by SynCAM 1. The Journal of Neuroscience. 2016;**36**(28):7464-7475

[88] Colucci-D'Amato L, Speranza L, Volpicelli F. Neurotrophic factor BDNF, physiological functions and therapeutic potential in depression, neuro-degeneration and brain cancer. International Journal of Molecular Sciences. 2020;**21**(20):p. 7777.

[89] Bartkowska K et al. Trk signaling regulates neural precursor cell proliferation and differentiation during cortical development. Development. 2007;**134**(24):4369-4380

[90] Lamb YN et al. Brain-derived neurotrophic factor Val66Met polymorphism, human memory, and synaptic neuroplasticity. Wiley Interdisciplinary Reviews: Cognitive Science. 2015;**6**(2):97-108

[91] Egan MF et al. The BDNF val66met polymorphism affects activity-dependent secretion of BDNF and human memory and hippocampal function. Cell. 2003;**112**(2):257-269

[92] Itami C et al. Brain-derived neurotrophic factor-dependent unmasking of "silent" synapses in the developing mouse barrel cortex. Proceedings of the National Academy of Sciences of the United States of America. 2003;**100**(22):13069-13074

[93] Edelmann E, Lessmann V, Brigadski T. Pre- and postsynaptic twists in BDNF secretion and action in synaptic plasticity. Neuropharmacology. 2014;**76**(Pt C):610-627

[94] Cechova K et al. Impact of APOE and BDNF Val66Met gene polymorphisms on cognitive functions in patients with amnestic mild cognitive impairment. Journal of Alzheimer's Disease. 2020;**73**(1):247-257

[95] Malykhin NV et al. The associations of the BDNF and APOE polymorphisms, hippocampal subfield volumes, and episodic memory performance across the lifespan. Hippocampus. 2020;**30**(10): 1081-1097

[96] Bambah-Mukku D et al. A positive autoregulatory BDNF feedback loop via C/EBPbeta mediates hippocampal memory consolidation. The Journal of Neuroscience. 2014;**34**(37):12547-12559

[97] Moutinho D, Veiga S, Requena JR. Human VGF-derived antidepressant neuropeptide TLQP62 promotes SH-SY5Y neurite outgrowth. Journal of Molecular Neuroscience. 2020;**70**(8): 1293-1302

[98] Jiang C et al. VGF function in depression and antidepressant efficacy. Molecular Psychiatry. 2018;**23**(7): 1632-1642

[99] Jiang C, Lin WJ, Salton SR. Role of a VGF/BDNF/TrkB autoregulatory feedback loop in rapid-acting antidepressant efficacy. Journal of Molecular Neuroscience. 2019;**68**(3): 504-509

[100] El Hayek L et al. Lactate mediates the effects of exercise on learning and memory through SIRT1-dependent activation of hippocampal brain-derived neurotrophic factor (BDNF). The Journal of Neuroscience. 2019; **39**(13):2369-2382

[101] Mohammadi M et al. Microglia dependent BDNF and proBDNF can impair spatial memory performance during persistent inflammatory pain. Behavioural Brain Research. 2020;**390**: 112683

[102] Tapia-Gonzalez S et al. Smad3 is required for the survival of proliferative intermediate progenitor cells in the dentate gyrus of adult mice. Cell Communication and Signaling: CCS. 2013;**11**:93

[103] Tapia-Gonzalez S et al. Dopamine and alpha-synuclein dysfunction in Smad3 null mice. Molecular Neurodegeneration. 2011;**6**:72

[104] Havranek T et al. Intracerebroventricular oxytocin administration in rats enhances object recognition and increases expression of neurotrophins, microtubule-associated protein 2, and synapsin I. Journal of Neuroscience Research. 2015;**93**(6): 893-901

[105] Lin YT, Hsu KS. Oxytocin receptor signaling in the hippocampus: Role in regulating neuronal excitability, network oscillatory activity, synaptic plasticity and social memory. Progress in Neurobiology. 2018;**171**:1-14

[106] Reichova A et al. Abnormal neuronal morphology and altered synaptic proteins are restored by oxytocin in autism-related SHANK3 deficient model. Molecular and Cellular Endocrinology. 2020;**518**:110924

[107] Baracz SJ et al. Maternal separation changes maternal care, anxiety-like behaviour and expression of paraventricular oxytocin and corticotrophin-releasing factor immunoreactivity in lactating rats. Journal of Neuroendocrinology. 2020;**32**(6):e12861

[108] Bakos J et al. Molecular mechanisms of oxytocin signaling at the synaptic connection. Neural Plasticity. 2018;**2018**:4864107

[109] Kozareva DA et al. Absence of the neurogenesis-dependent nuclear receptor TLX induces inflammation in the hippocampus. Journal of Neuroimmunology. 2019;**331**:87-96

[110] Burke AR et al. Impact of adolescent social experiences on behavior and neural circuits implicated in mental illnesses. Neuroscience and Biobehavioral Reviews. 2017;**76** (Pt B):280-300

[111] Haque FN et al. Social defeat interacts with Disc1 mutations in the mouse to affect behavior. Behavioural Brain Research. 2012;**233**(2):337-344

[112] Ibi D et al. Combined effect of neonatal immune activation and mutant DISC1 on phenotypic changes in adulthood. Behavioural Brain Research. 2010;**206**(1):32-37

[113] Nagai T et al. Effects of antipsychotics on the behavioral deficits in human dominant-negative DISC1 transgenic mice with neonatal polyI:C treatment. Behavioural Brain Research. 2011;**225**(1):305-310

[114] Li N et al. Adolescent isolation interacts with DISC1 point mutation to impair adult social memory and synaptic functions in the Hippocampus. Frontiers in Cellular Neuroscience. 2018;**12**:238

[115] Uzuneser TC et al. Disrupted-in-Schizophrenia 1 (DISC1) overexpression and juvenile immune activation cause sex-specific Schizophrenia-related psychopathology in rats. Frontiers in Psychiatry. 2019;**10**:222

Chapter 2

FAD-Linked Autofluorescence and Chemically-Evoked Zinc Changes at Hippocampal Mossy Fiber-CA3 Synapses

Fatima M.C. Bastos, Carlos M. Matias, Ines O. Lopes, João P. Vieira, Rosa M. Santos, Luis M. Rosario, Rosa M. Quinta-Ferreira and Maria Emilia Quinta-Ferreira

Abstract

Glutamatergic vesicles in hippocampal mossy fiber presynaptic boutons release zinc, which plays a modulatory role in synaptic activity and LTP. In this work, a fluorescence microscopy technique and the fluorescent probe for cytosolic zinc, Newport Green (NG), were applied, in a combined study of autofluorescence and zinc changes at the hippocampal mossy fiber-CA3 synaptic system. In particular, the dynamics of flavoprotein (FAD) autofluorescence signals, was compared to that of postsynaptic zinc signals, elicited both by high K^+ (20 mM) and by tetraethylammonium (TEA, 25 mM). The real zinc signals were obtained subtracting autofluorescence values, from corresponding total NG-fluorescence data. Both autofluorescence and zinc-related fluorescence were raised by high K+. In contrast, the same signals were reduced during TEA exposure. It is suggested that the initial outburst of TEA-evoked zinc release might activate ATP-sensitive K^+ (K_{ATP}) channels, as part of a safeguard mechanism against excessive glutamatergic action. This would cause sustained inhibition of zinc signals and a more reduced mitochondrial state. In favor of the "K_{ATP} channel hypothesis", the K_{ATP} channel blocker tolbutamide (250 μM) nearly suppressed the TEA-evoked fluorescence changes. It is concluded that recording autofluorescence from brain slices is essential for the accurate assessment of zinc signals and actions.

Keywords: autofluorescence, Newport Green, K_{ATP} channels, TEA, tolbutamide, hippocampal slices

1. Introduction

Glutamatergic vesicles in hippocampal mossy fiber presynaptic boutons sequester and actively release zinc together with glutamate [1, 2]. Once released, zinc diffuses in the synaptic cleft and binds to specific sites mainly in the postsynaptic CA3 neuronal membrane, crossing it via calcium-permeable glutamate receptors and voltage-dependent calcium channels (VDCCs) [3–8]. Thus, zinc plays significant modulatory roles in synaptic activity and possibly also in long-term potentiation

(LTP) [2, 9–12]. A negative feedback action at presynaptic sites seems also to occur, when released zinc activates the ATP-sensitive K^+ (K_{ATP}) channels [13, 14], besides inhibiting VDCCs [15, 16] on the boutons. These processes are thought to protect synapses from excessive neurotransmitter release and, consequently, too much postsynaptic activity.

In previous studies, performed using intracellularly trapped fluorescent dyes as zinc probes, it has been shown that single and tetanic stimulation of mossy fibers evoke zinc release and postsynaptic intracellular zinc rises [17–19]. It has also been reported that the application of high external potassium concentrations elicits a strong depolarization in the cells [20, 21], as well as postsynaptic zinc enhancements [22]. However, the application of intense tetanic stimulation caused a depression of zinc and glutamate release, which was reduced by the application of the K_{ATP} channel blocker tolbutamide [14]. Furthermore, in the presence of the potassium channel blocker tetraethylammonium (TEA), which induces a weaker depolarization in the mossy fiber region [23], the zinc signals were reversibly depressed being this depression also reduced by tolbutamide [24]. Since the depolarizing effect of KCl is stronger than that of TEA, it is expected that the increase of the synaptic activity will be more intense in the presence of KCl. The postsynaptic actions of zinc may not be limited to the membrane or the cytosol, as there is evidence that this cation is taken up by cell organelles including the endoplasmic reticulum and mitochondria [4, 25, 26]. Recording zinc probe fluorescence from slices is not devoid of artifacts. A potential problem is the tissue autofluorescence, which depends on the excitation wavelengths, which may arise from both NAD(P)H (near-UV) and flavoprotein-bound FAD (visible) [27–29]. In the present experiments, which were performed using the fluorescent zinc indicator Newport Green (NG), the prevailing contributor to autofluorescence was the FAD-linked fluorescence, since NG was excited with visible light. The redox couple FAD/$FADH_2$ operates in the citric acid cycle and respiratory chain, and this autofluorescence component might originate mostly, if not exclusively from mitochondria [30–32]. Furthermore, dehydrogenases are calcium-sensitive [33], suggesting that FAD-linked autofluorescence might change following stimulation of hippocampal mossy fibers. Thus, an important objective of this study was to determine whether mossy fiber autofluorescence might affect the zinc signals and, if so, to extract the real dynamics of the zinc signals from the total fluorescence recordings. Autofluorescence of mitochondrial origin reflects the metabolic activity of the cells. Therefore, carrying out parallel autofluorescence recordings offers the possibility of correlating the postsynaptic zinc dynamics of pyramidal CA3 neurons with changes in tissue metabolic activity.

2. Experimental procedures

All experiments were carried out in accordance with the Directive 2010/63/EU of the European Parliament and Council. In agreement with the Portuguese law, the animal care review committee is the *Direção- Geral de Alimentação e Veterinária (DGAV)*. All efforts were made to minimize animal suffering and to use only the number of animals necessary to produce reliable scientific data.

The experiments were performed in the synaptic system mossy fibers-CA3 pyramidal cells, using the brain of pregnant Wistar rat females (16–18 weeks old), kindly provided by the CNC (Center for Neurosciences) laboratory (as stated in the acknowledgments). The hippocampal slices were prepared as previously described [22, 24]. Briefly, after cervical dislocation, the isolated brain was rapidly cooled (5–8°C) in oxygenated (95% O_2, 5% CO_2) artificial cerebrospinal fluid (ACSF), containing (in mM): NaCl 124; KCl 3.5; $NaHCO_3$ 24; NaH_2PO_4 1.25; $MgCl_2$ 2; $CaCl_2$

2 and D-glucose 10, pH 7.4. Then, the hippocampus was separated and the slices (400 μm thick) were cut transversely and immersed in ACSF at room temperature. After a resting period of about 1 h, the slices were transferred to the experimental chamber and perfused with ACSF, at a rate of 1.5 to 2 ml/min, T = 30–32°C.

Autofluorescence measurements were carried out in slices immersed in ACSF or in the medium of interest. For the detection of zinc changes, the slices were incubated during one hour in an ACSF solution containing 5 μM of the permeant form of the fluorescent zinc indicator Newport Green (NG). The preparation of the solution was performed as follows: 1 mg of NG was dissolved in 250 μl of DMSO and then 5 μl of this mixture (DMSO + NG) were diluted in 5 ml of ACSF containing 5 μl of pluronic acid F-127. The indicator NG has a moderate affinity for zinc ($K_d \sim 1$ μM) and a relatively low affinity for calcium ($K_d > 100$ mM).

The fluorescence signals recorded from NG loaded slices contain an autofluorescence and a zinc-NG components. Having been realized that the first component varies both with time and with the perfusing medium, fluorescence signals from dye-free slices were recorded using similar experimental protocols as those applied in the NG containing slices. The real zinc signals were thus obtained from those measured from the indicator-loaded slices, after point-to-point subtraction of the intrinsic fluorescence records. Both types of records were normalized by the average of the baseline (first ten points) values prior to the mentioned subtraction, to avoid non-intrinsic and non-dye associated fluorescence. For the correction procedure, the following equations were used:

$$F_{A0} = \frac{\sum_{i=1}^{10} F_{Ai}}{10} \tag{1}$$

The corrected and normalized signals were represented as:

$$\frac{F_i}{F_0} = 1 + \left(\frac{F_{Ti}}{F_{T0}} - \frac{F_{Ai}}{F_{A0}} \right) \tag{2}$$

F_{A0} – basal autofluorescence; F_{Ai} – autofluorescence; F_{T0} – basal total fluorescence; F_{Ti} – total fluorescence; F_i -. zinc fluorescence; F_0 – basal zinc fluorescence.

In order to study the role of zinc in autofluorescence changes evoked by chemical stimulation with KCl, the zinc chelator ethylenediaminetetraaceticacid disodium calcium salt (Ca-EDTA), (25 mM), was, in some experiments, added to the medium. The KCl solution consisted of ACSF but with 20 mM concentration of KCl. The tetraethylammonium (TEA) solution consisted of ACSF with higher concentrations of $CaCl_2$ and KCl, 10 mM and 5 mM, respectively, and with TEA at a concentration of 25 mM. The K_{ATP} channel blocker tolbutamide (250 μM) was directly applied to the perfusion medium, which was recirculated. All solutions were applied for periods of 30 min, except the KCl + Ca-EDTA solution that was perfused for 1 h.

The detection and measurement of zinc signals was performed using an experimental transfluorescence setup based on a microscope (Zeiss Axioskop), equipped with a halogen light source (12 V, 100 W). Excitation and emission wavelengths were selected by means of a narrowband filter (480 nm, BW 10 nm) and a high pass filter (> 500 nm), respectively. The capture of the transmitted light was done by a water immersion lens (40x, N.A. 0.75), being that light focused on a photodiode (Hammamatsu, 1 mm^2). Subsequently, the current of the photodiode was converted in an electrical signal by a current/voltage converter (I/V), with a feedback resistance of 1 GΩ. The signal was digitally processed by a 16 bit analog/digital converter

(f = 1.67 Hz) and analyzed using the Signal Express™ software from National Instruments. The average value of each group of 100 consecutive points was used in order to illustrate the results.

All data are expressed as mean ± SEM. The results of the statistical studies were obtained using the Mann–Whitney U test ($p < 0.05$).

In the experiments, the chemical products used were: NG, Pluronic acid F-127 (Life technologies, Carlsbad, CA); DMSO, TEA, Tolbutamide (Sigma-Aldrich, Sintra, PT).

3. Results

In this study, fluorescence changes evoked by the application of KCl and TEA media were measured both from non-incubated and from NG loaded slices. In most figures the first 10 min represent data from slices exposed to the ACSF solution, the next 30 min data evoked by the media of interest (KCl or TEA) and in the following 30 min ACSF was again perfused. Since the excitation wavelength used was 480 nm, the recorded fluorescence emission from dye-free slices is considered essentially autofluorescence, with flavoprotein origin. Taking into account the spectral characteristics of the intrinsic fluorescence of FAD and of zinc-bound NG, the fluorescence signals obtained from slices incubated in NG, using excitation light of 480 nm and recording above 500 nm, have two different components: autofluorescence and the fluorescence of the zinc-NG complex. The results of a group of experiments designed to extract the real zinc signal, as the difference between those two components, are shown in **Figure 1**.

The autofluorescence trace (**Figure 1a**) reveals that the KCl (20 mM) solution caused an increase of the signals, of 11 ± 2% in the period 35–40 min (n = 4). The total fluorescence signals (**Figure 1b**) were enhanced by 36 ± 5% above the basal values in the same period (n = 3). Thus, it can be observed that the zinc signals (**Figure 1c**), obtained as the total fluorescence minus the autofluorescence changes, reached a stable value having an amplitude of 27 ± 3% during the last 5 min in KCl. The autofluorescence changes continued to increase during the reperfusion of ACSF, reaching a maximum between 45 and 50 min after the beginning of the experiment, corresponding to an amplitude of 20 ± 2% above baseline. Afterwards they decreased again to 11 ± 2%. The total fluorescence trace decreased less during washout, having an amplitude of 29 ± 4% during the last 5 min. The zinc signals decreased during the first 10 min of washout but were maintained in the remaining period at 16 ± 4% above baseline.

Unlike the behavior observed with KCl, both the autofluorescence and the total fluorescence signals were reduced in the presence of 25 mM TEA, recovering, upon its removal, to a value above the baseline (**Figure 2a** and **b**). The application of TEA caused a decline of the amplitude of the total signal (**Figure 2b**) by 17 ± 3% of control (35–40 min., n = 8). In these experiments, autofluorescence (**Figure 2a**) was responsible for almost half of the depression since the intrinsic signals decreased by 8 ± 2% (35–40 min, n = 5). Consequently, it can be concluded that the zinc signals (**Figure 2c**), obtained again as the difference between the traces in panels b and a, were reduced by 9 ± 1%, in the same period.

The superimposed results of the autofluorescence and of the corresponding zinc signals are shown in **Figure 3**, for the experiments performed with KCl (**Figure 3a**) and TEA (**Figure 3b**). They reveal that the signals from the unincubated and the NG-treated slices have different time courses in the case of KCl, rising the autofluorescence transients more slowly than the zinc ones (**Figure 3a**). TEA causes roughly similar autofluorescence and zinc changes (**Figure 3b**). It should also be noticed that the KCl evoked zinc changes remained potentiated upon washout, while, following TEA removal the zinc signals recovered completely.

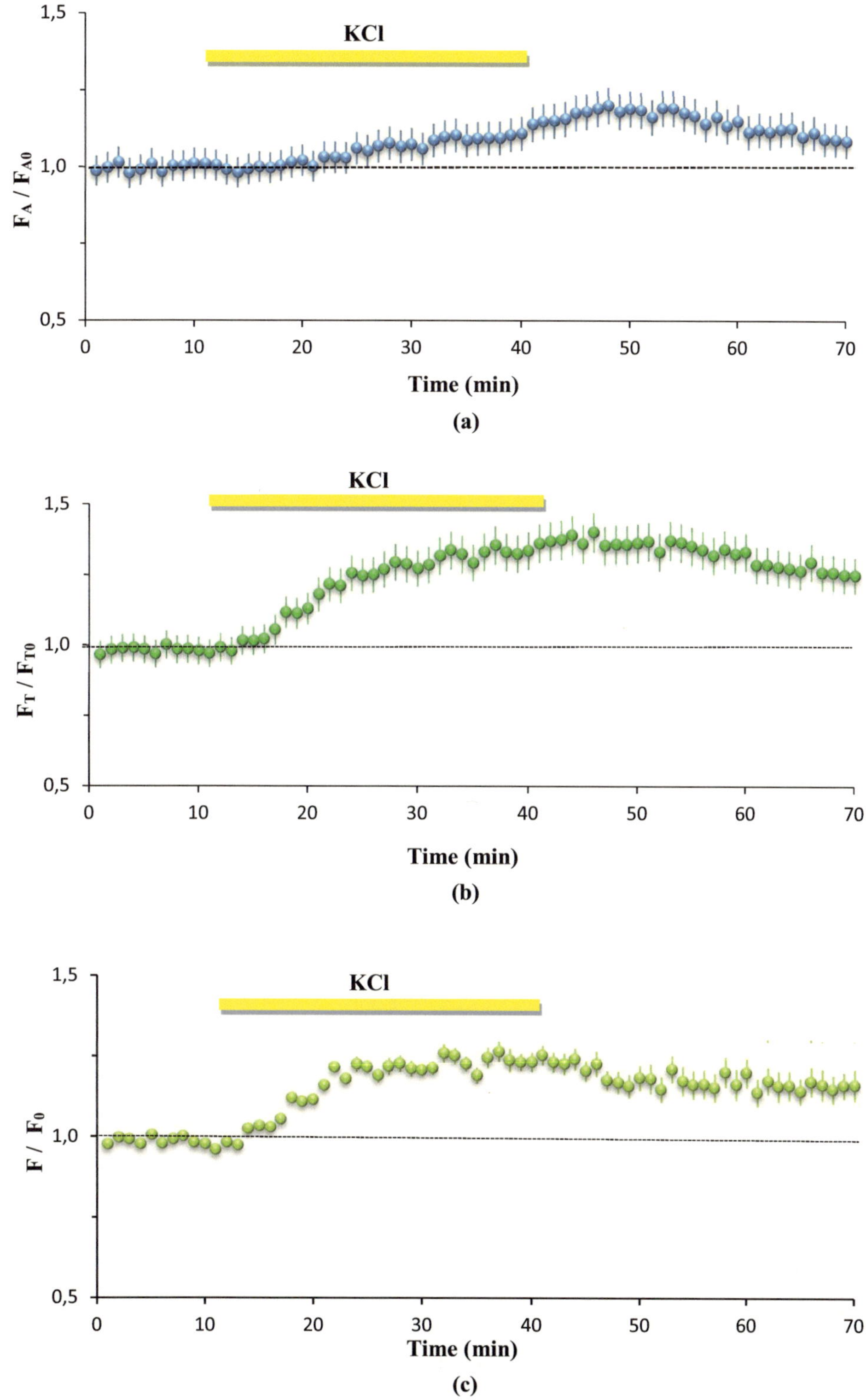

Figure 1.

Autofluorescence, total fluorescence and zinc signals evoked by KCl. (a) KCl-induced changes in autofluorescence (n = 4). (b) Total fluorescence data obtained in slices incubated with NG (n = 3). (c) Zinc signals obtained as the difference between data in panels (b) and (a). The application of KCl (20 mM) was made at the times indicated by the bars. All values were normalized by the average of the first 10 responses and represent the mean ± SEM. F_A, autofluorescence; F_T, total fluorescence; F, zinc signals; F_{A0}, F_{T0}, F_0, basal auto-, total and zinc fluorescences, respectively.

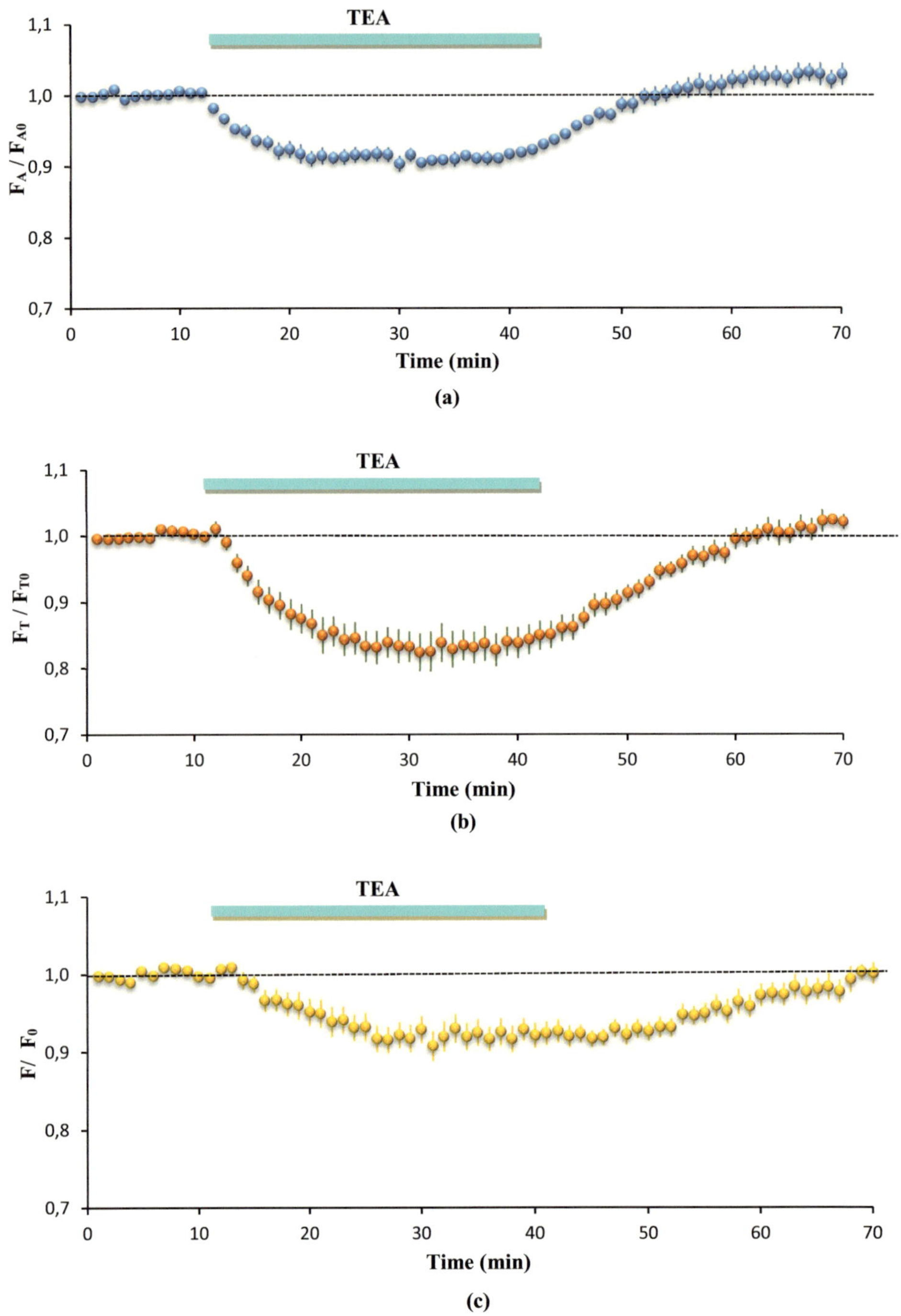

Figure 2.

Pooled data of autofluorescence, total fluorescence and zinc signals evoked by TEA. (a) Effect of the application of TEA (25 mM) on autofluorescence changes (n = 5). (b) Total fluorescence signals obtained from slices incubated with NG (n = 8). (c) Zinc signals given by the difference between the traces in panels (a) and (b). The solution with TEA (25 mM) was perfused during the period indicated by the bars. All values were normalized by the average of the first 10 responses and represent the mean ± SEM. F_A, autofluorescence; F_T, total fluorescence; F, zinc signals; F_{A0}, F_{T0}, F_0, basal auto-, total and zinc fluorescences, respectively.

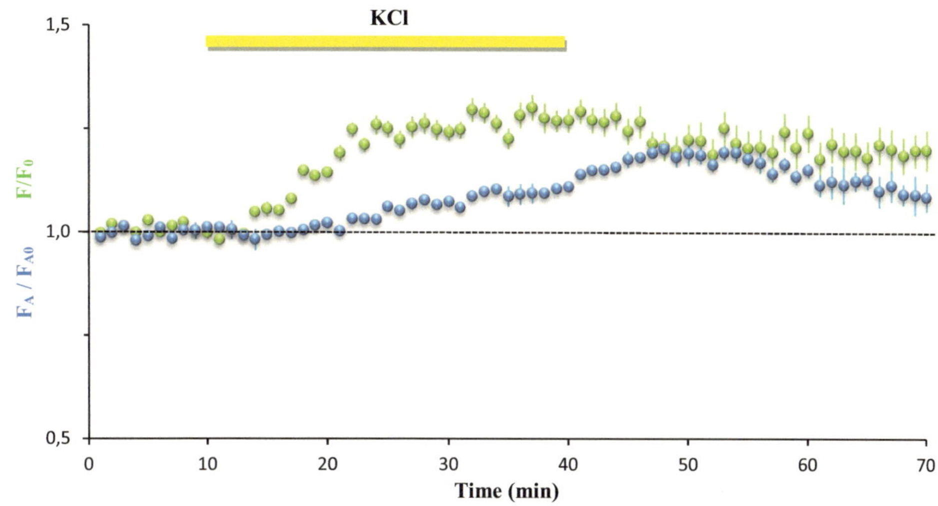

(a)

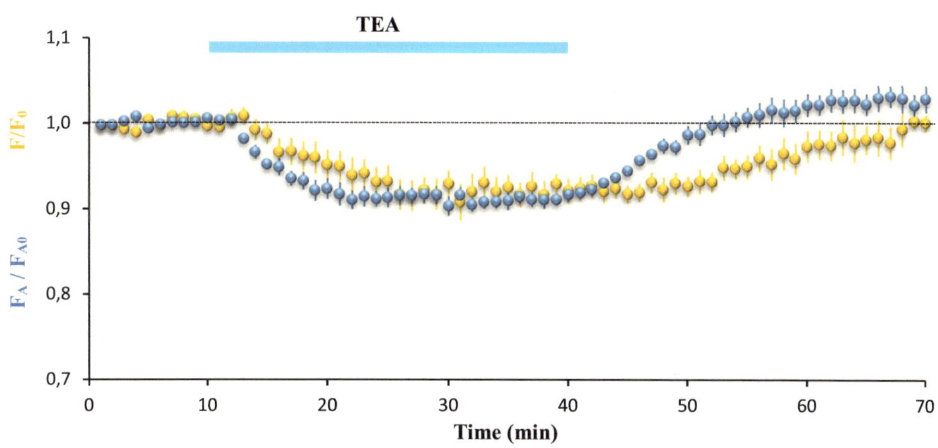

(b)

Figure 3.
Superimposed curves of the KCl- and TEA-evoked fluorescence signals. (a) Autofluorescence (blue, n = 3) and zinc (green) signals induced by KCl (20 mM). (b) Autofluorescence (blue, n = 5) and zinc (orange) changes evoked by TEA (25 mM). The solutions of KCl and TEA were applied at the times indicated by the bars. All values were normalized by the average of the first 10 responses and represent the mean ± SEM. F_A, autofluorescence; F, zinc fluorescence; F_{Ao}, F_o, basal auto- and zinc fluorescences.

In order to verify the involvement of K_{ATP} channels in the TEA induced fluorescence depression, tolbutamide, a K_{ATP} channel blocker was applied with TEA, at the concentrations of 250 μM and 25 mM, respectively. In the control group of experiments, the TEA solution was perfused twice, circulating ACSF for 30 min after each TEA application, in order to determine the degree of recovery upon washout. In the other group of experiments, the second TEA solution contained also tolbutamide.

According to the results in **Figure 4**, both perfusions of the TEA solution caused similar declines in the intensity of the autofluorescence signals, which have also similar time courses as illustrated in **Figure 4a**. The first depression had an amplitude of 11 ± 1%, with respect to baseline (35–40 min, n = 3). The second one had an

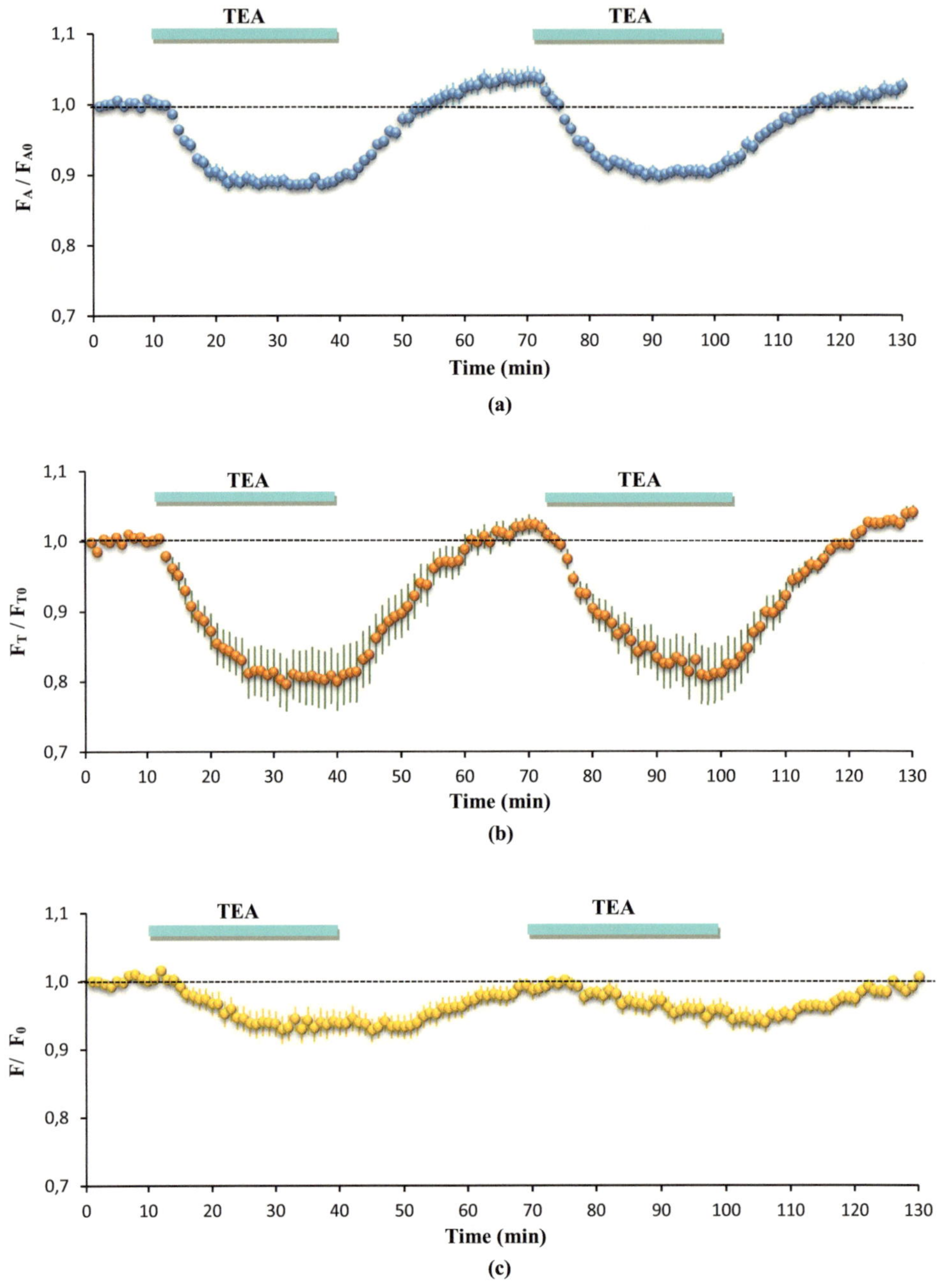

Figure 4.

Fluorescence signals induced by two consecutive applications of TEA. (a) Autofluorescence responses (n = 3). (b) Total fluorescence data (n = 8). (c) Zinc signals calculated subtracting trace (a) from trace (b). The solution with TEA (25 mM) was circulated during the period indicated by the bars. All values were normalized by the average of the first 10 responses and are represented as mean ± SEM. F_A, autofluorescence; F_T, total fluorescence; F, zinc signals; F_{A0}, F_{T0}, F_0, basal, auto-, total and zinc fluorescences, respectively.

amplitude of 10 ± 2% (95–100 min, n = 3). The washout of the TEA solution was, as expected, accompanied by the recovery of the autofluorescence signal, which reached, at the end of the subsequent 30 min period in ACSF, a small potentiation, 3 ± 1% at 65–70 min and 2 ± 1% at 125–130 min.

In a similar type of experiments, the second TEA application was combined with tolbutamide (250 μM) (**Figure 5**). This compound had no effect on the

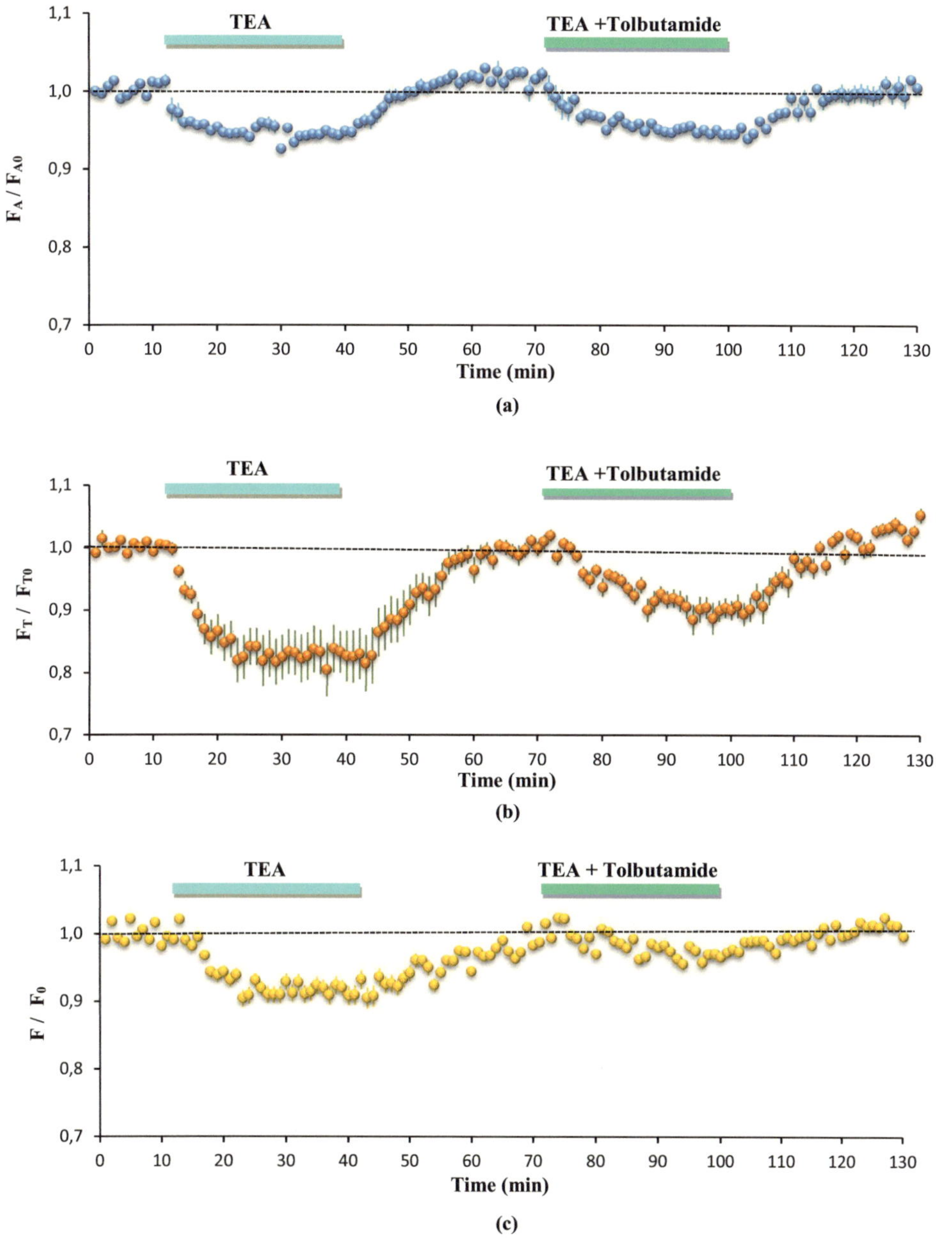

Figure 5.
Tolbutamide significantly blocks zinc signals evoked by TEA. (a) Effect of the application of TEA (25 mM) followed by TEA (25 mM) plus tolbutamide (250 μM) on autofluorescence (n = 2). (b) Similar records for total fluorescence (n = 3). (c) Zinc signals determined as trace (b) minus (a). The solutions of TEA and TEA plus tolbutamide were applied at the times indicated by the bars. The data were normalized by the average of the first 10 responses and represent the mean ± SEM. F_A, autofluorescence; F_T, total fluorescence; F, zinc signals; F_{A0}, F_{T0}, F_0, basal auto-, total and zinc fluorescences, respectively.

autofluorescence signal since the second depression was similar to the first one (**Figure 5a**), as observed in the experiment with two consecutive TEA applications (**Figure 4a**). On the other hand, the application of tolbutamide significantly blocked the depression of the total (**Figure 5b**) and of the zinc (**Figure 5c**) signals, having the latter been, once more, obtained subtracting the autofluorescence component from the total fluorescence changes. As noticed before, tolbutamide applied only in ACSF had essentially no effect on the autofluorescence depression.

The application of TEA evoked a depression of the zinc signal with an amplitude of 8 ± 1% (35–40 min, n = 3), while in the presence of tolbutamide, the corresponding amplitude was only 3 ± 1% (35–40 min, n = 3). Thus, tolbutamide blocks about two thirds of the TEA evoked zinc signal inhibition, suggesting that the zinc, but not the autofluorescence depression, is mainly mediated by the activation of K_{ATP} channels.

4. Discussion

The experiments presented in this work, performed at the hippocampal mossy fiber synapses from CA3 area, allow the comparison of autofluorescence signals, recorded from non-incubated slices, with fluorescence zinc changes obtained from NG-loaded slices, after subtraction of the autofluorescence component. At these synapses, it

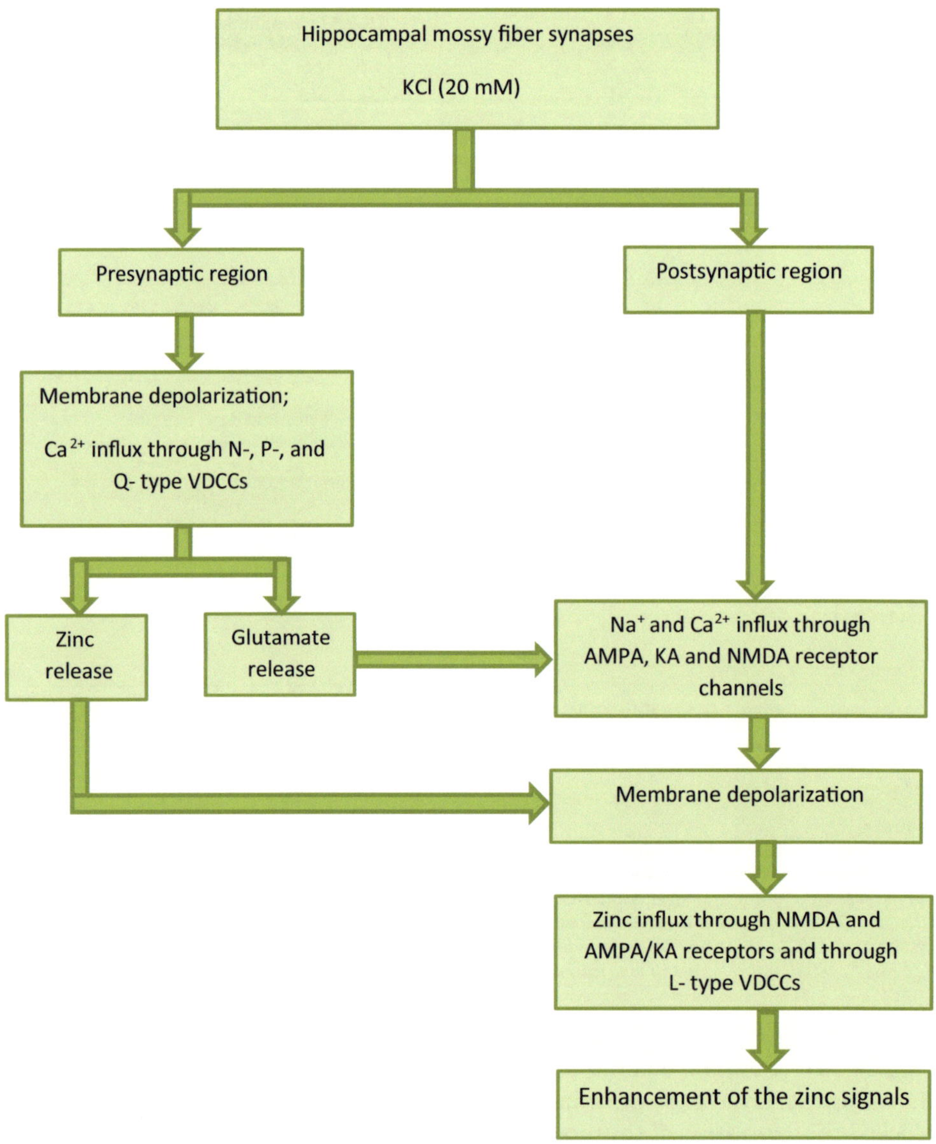

Figure 6.
Schematic representation of the major cellular mechanisms and ionic fluxes involved in KCl depolarization (20 mM) at the hippocampal mossy fibers-CA3 pyramidal cells synapses.

was observed that chemically induced depolarization by KCl (20 mM), evoked clear autofluorescence changes that recovered partially during the 30 min period following washout. These changes have a lower amplitude and a slower time course than those of the similarly evoked fluorescence signals observed in slices incubated with the zinc indicator Newport Green. The latter signals, corresponding to the total fluorescence changes in this work, have previously been described by Bastos et al. [22]. The application of 20 mM KCl promotes the depolarization of the presynaptic membrane, mediated by voltage dependent potassium channels (VDKCs), to a resting value of about −54 mV [13]. This increase in the membrane potential activates presynaptic VDCCs, triggering glutamate and zinc co-release, as illustrated in **Figure 6**.

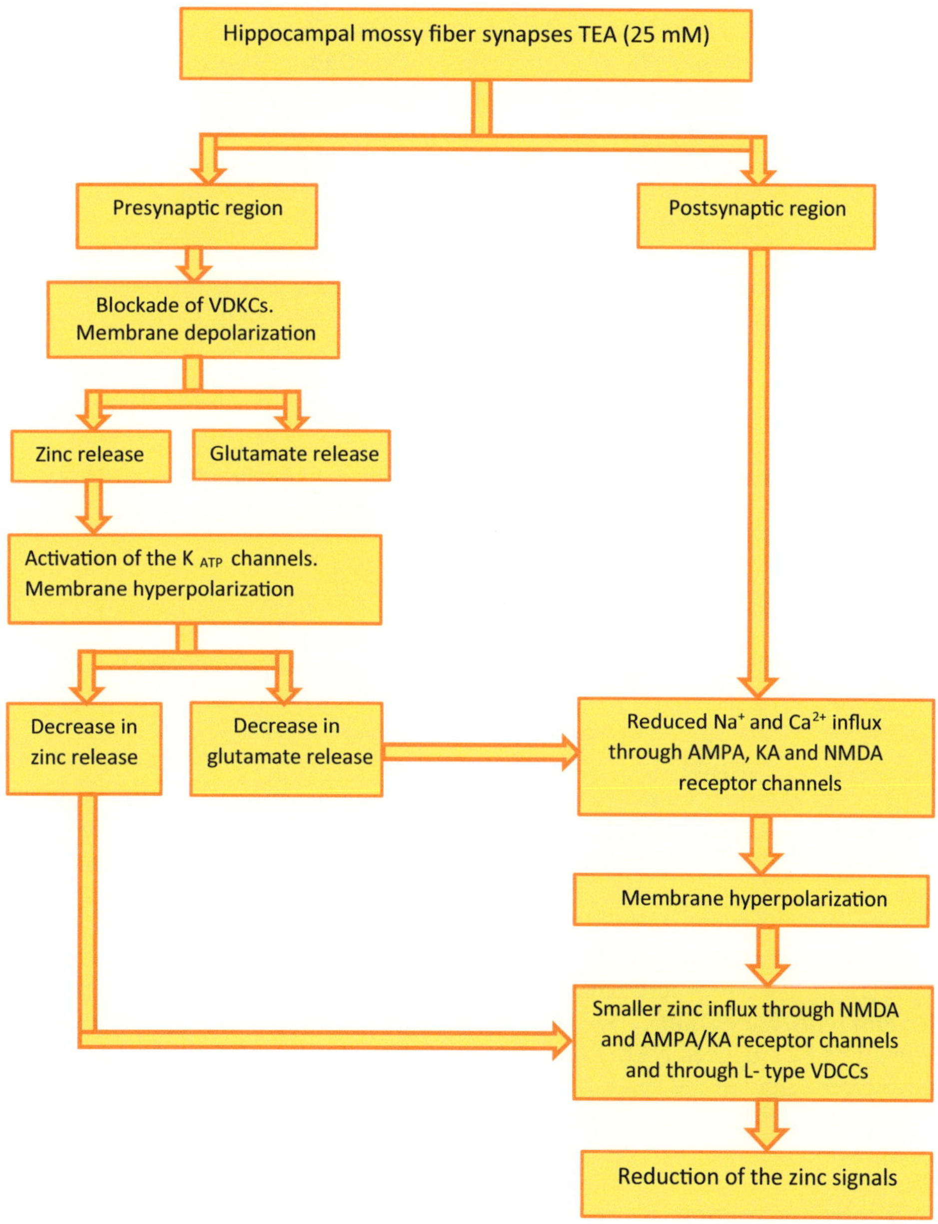

Figure 7.
Diagram of the principal mechanisms and ionic movements involved in TEA depolarization at the hippocampal mossy fiber-CA3 synaptic system.

Subsequently, after diffusion in the cleft and binding to specific pre- and postsynaptic sites, zinc flows into the postsynaptic region through several channels and receptors, including NMDA and calcium permeable AMPA/KA receptors and L- and T-type VDCCs [3–5, 8, 34]. There is also experimental evidence that zinc can be released from intracellular sources following blockade of ERs [35, 36]. Consequently, at the postsynaptic region, calcium and zinc entry through both glutamate receptor channels and VDCCs leads to cytosolic calcium and zinc accumulation that may cause the flow of both ions to mitochondria, through the activation of the mitochondrial Ca-uniporter [30, 37–43].

TEA also causes membrane depolarization that may trigger various cellular processes, as described in **Figure 7**.

In this work, the autofluorescence changes evoked by a single or by consecutive identical TEA applications were all similar. Interestingly, the time course of these changes is similar to that of the corresponding fluorescence signals recorded from Newport Green loaded slices. The TEA triggered fluorescence changes are depressed during the perfusion of TEA and recover to or above the baseline level upon washout, as previously reported [24]. In the present study, after subtracting the autofluorescence component, the real zinc signals evoked by TEA have about half the amplitude of that of the total fluorescence traces.

The changes of both autofluorescence and total fluorescence TEA induced signals are the opposite of those evoked by KCl. As previously mentioned, the membrane depolarization is higher in the presence of KCl than in TEA. The blockade of presynaptic VDKCs by TEA evokes a weak depolarization [23], followed by glutamate and zinc release and the activation of K_{ATP} channels by this ion. This leads to membrane hyperpolarization with a lower amplitude than that of the KCl evoked depolarization. Consequently, the hyperpolarizing effect of the zinc induced activation of presynaptic K_{ATP} channels can be occulted by the large increase in the resting potential, due to the strong KCl evoked depolarization, mediated by VDKCs [13].

5. Conclusions

The amount of calcium entry is related to the intensity of autofluorescence because increased intracellular calcium and zinc can trigger an increase in FAD (flavoprotein) and NAD, as well as in the oxidation of $FADH_2$ and NADH [28, 44]. In the present experimental conditions (excitation wavelength of 480 nm, emission light collected above 500 nm) and taking into account the spectral properties of FAD, the autofluorescence detected is considered to have FAD origin. As previously mentioned, KCl depolarization causes the entry of both calcium and zinc ions to the postsynaptic region [10, 22, 24, 45, 46]. When increases in cytosolic zinc concentration are high, zinc ions enter the mitochondria, and if in excess may have neurotoxic effects [4, 47, 48]. Thus, the origin of the KCl evoked autofluorescence signals is, under our experimental conditions and based on previous studies, the flavoproteins. For example, Pancani [49], have found, using the complex I inhibitor rotenone a KCl induced NADH fluorescence decrease of mitochondrial origin. This is in agreement with the observed FAD enhancement since the NADH and FAD fluorescence changes are opposite [29, 44, 50].

Acknowledgements

We thank CNC - Center for Neuroscience and Cell Biology, University of Coimbra, Coimbra, Portugal, for providing the rat brains. Work funded by strategic project UID/NEU/04539/2013. All authors are aware of and have approved the manuscript as submitted. All authors are aware they must each submit a completed License to Publish form before the manuscript can be published. (Each author will receive an email with instructions).

Author details

Fatima M.C. Bastos[1,2], Carlos M. Matias[1,3*], Ines O. Lopes[2], João P. Vieira[4], Rosa M. Santos[1,4], Luis M. Rosario[1,4], Rosa M. Quinta-Ferreira[5] and Maria Emilia Quinta-Ferreira[1,2]

1 CNC, University of Coimbra, Coimbra, Portugal

2 Department of Physics, University of Coimbra, Coimbra, Portugal

3 Department of Physics, School of Science and Technology, UTAD, Vila Real, Portugal

4 Department of Life Sciences, University of Coimbra, Coimbra, Portugal

5 CIEPQPF-Department of Chemical Engineering, University of Coimbra, Coimbra, Portugal

*Address all correspondence to: cmatias@utad.pt

References

[1] Vallee, B.L. and Falchuk, K.H. (1993). The biochemical basis of zinc physiology. Physiol. Rev. 73: 79-118. https:/doi.org/10.1152/physrev.1993.73.1.7 PMID: 8419966.

[2] Frederickson C.J., Suh S.W., Silva D., Frederickson C.J. and Thompson R.B. (2000). Importance of zinc in the central nervous system: The zinc-containing neuron. J. Nutr. 130: 1471S–1483S.https://doi.org/10.1093/jn/130.51471S PMID: 10801962.

[3] Sensi S.L., Canzoniero L.M.T., Yu S.P., Ying H.S., Koh J-Y., Kerchner G.A. and Choi D.W. (1997). Measurement of intracellular free zinc in living cortical neurons, routes of entry. J. Neurosci. 17: 9554-9564.https://doi.org/10.1523/JNEUROSCI.17-24-09554.1997 PMID: 9391010.

[4] Sensi S.L., Yin H.Z., Carriedo S.G., Rao S.S. and Weiss J.H. (1999). Preferential Zn^{2+} influx through Ca2+–permeable AMPA/ kainate channels triggers prolonged mitochondrial superoxide production. Proc. Natl. Acad. Sci. USA. 96: 2414-2419. https://doi.org/10.1073/pnas.96.5.2414 PMID: 10051656.

[5] Marin, P., Israël, M., Glowinski, J. and Prémont, J. (2000). Route of zinc entry in mouse cortical neurons, role in zinc-induced neurotoxicity. Eur. J. Neurosci. 12: 8-18.https://doi.org/10.1046/j.1460-9568.2000.00875. x PMID: 10651855.

[6] Dietz, R.M., Weiss, J.H. and Shuttleworth, C.W. (2008). Zn2+ influx is critical for some forms of spreading depression in brain slices. J. Neurosci. 28: 8014-8024. https://doi.org/10.1523/JNEUROSCI.0765-08. 2008 PMID: 18685026.

[7] Quinta-Ferreira, M.E, Sampaio dos Aidos, F.D.S., Matias, C.M., Mendes, P.J., Dionísio, J.C., Santos, R.M., Rosário, L.M. and Quinta-Ferreira, R.M. (2016). Modelling zinc changes at the hippocampal mossy fiber synaptic cleft. J. Comput. Neurosci. 41, 323-337 (2016). https:// doi.org/ 10.1007/s 10827-016-0620-x.

[8] Freitas, J.C.S, Miraldo, J.N., Matias, C., Sampaio Dos Aidos, F.D.S., Mendes, P.J., Dionisio, J.C., Santos, R.M., Rosario, L.M., Quinta-Ferreira, R.M. and Quinta-Ferreira M.E. (2019). Computer Simulations of Hippocampal Mossy Fiber Cleft Zinc Movements. Advances in Neural Signal Processing (ISBN 978-1-78984-114-5).

[9] Vogt, K., Mellor, J., Tong, G. and Nicoll, R. (2000). The actions of synaptically released zinc at hippocampal mossy fiber synapses. Neuron. 26: 187-196. https://doi.org/10.1016/S0896-6273 (00)81149-6.

[10] Li, Y., Hough, C.J., Frederickson, C.J. and Sarvey, J.M. (2001b). Rapid translocation of Zn2+ from presynaptic terminals into postsynaptic hippocampal neurons after physiological stimulation. J. Neurophysiol. 86: 2597-2604. https://doi.org/10.1152/jn.2001.86.5.2597 PMID: 11698545.

[11] Quinta-Ferreira, M.E. and Matias, C.M. (2005). Tetanically released zinc inhibits hippocampal mossy fiber calcium, zinc and postsynaptic responses. Brain Res. 1047: 1-9. https://doi.org/10.1016/j.brainres.2005.04.006 PMID: 15950598.

[12] Takeda A., Fuke S., Tsutsumi W. and Oku N. (2007). Negative modulation of presynaptic activity by zinc released from Schaffer collaterals. J. Neurosci. Res. 85: 3666-3672.https://doi.org/10.1002/jnr.21449 PMID: 17680671.

[13] Bancila, V., Nikonenko, I., Dunant, Y. and Bloc, A. (2004). Zinc inhibits

glutamate release via activation of pre-synaptic KATP channels and reduces ischaemic damage in rat hippocampus. J. Neurochem. 90: 1243-1250 https://doi.org/10.1111/j.1471-4159.2004.02587.x. PMID: 15312179.

[14] Matias, C.M., Saggau, P. and Quinta-Ferreira, M.E. (2010). Blockade of presynaptic KATP channels reduces the zinc-mediated posttetanic depression at hippocampal mossy fiber synapses. Brain Res. 1320: 22-27. https://doi.org/10.1016/j.brainres.2010.01.021 PMID: 20097182.

[15] Büsselberg, D., Michael, D., Evans, M.L., Carpenter, D.O. and Haas, H.L. (1992). Zinc (Zn2+) blocks voltage gated calcium channels in cultured rat dorsal root ganglion cells. Brain Res. 593: 77-81. http://doi.org/10.1016/0006-8993(92)91266-H PMID: 1333873.

[16] Harrison, N.L. and Gibbons, S.J. (1994). Zn2+: an endogenous modulator of ligand- and voltage-gated ion channels. Neuropharmacology. 33: 935-952.http://doi.org/10.1016/0028-3908(94)90152-x PMID: 7845550.

[17] Assaf, S.Y. and Chung, S.H. (1984). Release of endogenous Zn^{2+} from brain tissue during activity. Nature. 308: 734-736. https://doi.org/10.1006/exnr.1998.6931 PMID: 6717566.

[18] Budde, T., Minta, A., White, J. A. and Kay, A.R. (1997). Imaging free zinc in synaptic terminals in live hippocampal slices. Neurosci. 79: 347-358. https://doi.org/10.1016/S0306-4522(96)00695-1 PMID: 9200720.

[19] Quinta-Ferreira, M.E., Matias, C.M., Arif, M. and Dionísio, J.C. (2004). Measurement of presynaptic zinc changes in hippocampal mossy fibers. Brain Res. 1026: 1-10. https://doi.org/10.1016/j.brainres.2004.07.054. PMID: 15476692.

[20] Bernard, J., Lahsaini, A. and Massicotte, G. (1994). Potassium-induced long-term potentiation in area CA1 of the hippocampus involves phospholipase activation. Hippocampus. 4: 447-453. https://doi.org/10.1002/hipo.450040407 PMID: 7874236.

[21] Roisin, M.-P., Leinekugel, X. and Tremblay, E. (1997). Implication of protein kinase C in mechanisms of potassium- induced long-term potentiation in rat hippocampal slices. Brain Res. 745: 222-230. https://doi.org/10.1016/S0006-8993 (96) 01155-9.

[22] Bastos, F.M.C., Lopes, S.A., Corceiro, V.N., Matias, C.M., Dionísio, J.C., Aidos, F.D.S.S., Mendes P.J., Santos, R.M., Quinta-Ferreira, R.M. and Quinta-Ferreira, M.E. (2017a). Postsynaptic zinc potentiation elicited by KCl depolarization at hippocampal mossy fiber synapses. Gen. Physiol. Biophys. 36: 289-296. https://doi.org/10.4149/gpb_2017001 PMID: 28471347.

[23] Song, D., Wang, Z. and Berge, T.W. (2002). Contribution of T-Type VDCC to TEA-Induced Long-Term Synaptic Modification in Hippocampal CA1 and Dentate Gyrus. Hippocampus 12: 689 – 697. DOI 10.1002/hipo.10105.

[24] Bastos, F.M., Corceiro, V.N., Lopes, S.A., Almeida, J.G., Matias, C.M., Dionisio, J.C., Mendes, P.J., Sampaio dos Aidos, F.D.S., Quinta-Ferreira, R.M. and Quinta-Ferreira, M.E (2017b). Effect of tolbutamide on tetraethylammonium-induced postsynaptic zinc signals at hippocampal mossy fiber-CA3 synapses. Can. J. Physiol. Pharmacol. 95 (9): 1058-1063. https://doi.org/10.1139/cjpp-2016-0379 PMID: 28654763.

[25] Bossy-Wetzel, E., Talantova, M.V., Lee, W.D., Scholzke, M.N., Harrop, A., Mathews, E., Gotz, T., Han, J., Ellisman, M.H., Perkins, G.A. and Lipton, S.A. (2004). Crosstalk between nitric oxide

and zinc pathways to neuronal cell death involving mitochondrial dysfunction and p38-activated K^+ channels. Neuron. 41: 351-365.PMID: 14766175.

[26] Medvedeva, Y.V., Lin, B., Shuttleworth, C.W. and Weiss, J.H. (2009). Intracellular Zn^{2+} accumulation contributes to synaptic failure, mitochondrial depolarization, and cell death in an acute slice oxygen-glucose deprivation model of ischemia. J Neurosci. 29: 1105-1114. https://doi.org/10.1523/JNEUROSCI.4604-08.2009 PMID: 19176819.

[27] Reinert, K.C., Dunbar, R.L., Gao, W., Chen, G. and Ebner, T.J. (2004). Flavoprotein autofluorescence imaging of neuronal activation in the cerebellar cortex in vivo. J. Neurophysiol. 92(1), 199-211. https://doi.org/10.1152/jn.01275.2003 PMID: 14985415.

[28] Brennan, A.M.. Connor, J.A. and Shuttleworth, C.W. (2006). NAD(P)H Fluorescence Transients after Synaptic Activity in Brain Slices: Predominant Role of Mitochondria Function. J. Cereb. Blood Flow Metab. 26: 1389-1406.https://doi.org/10.1038/sj.jcbfm.9600292.

[29] Shuttleworth, C.W. (2010). Use of NAD(P)H and Flavoprotein Autofluorescence Transients to Probe Neuron and Astrocyte Responses to Synaptic Activation. Neurochem. Int. 56: 379-386. https://doi.org/10.1016/j.neuint.2009.12.015 PMID: 20036704.

[30] Sensi, S. L., Ton-That, D., Sullivan, P.G., Jonas, E.A., Gee, K.R., Kaczmarek, L.K. and Weiss, J.H. (2003). Modulation of mitochondrial function by endogenous Zn^{2+} pools. PNAS. 10: 6157-6162. http://doi/10.1073/pnas.1031598100 PMID: 12724524.

[31] Sensi, S.L., Paoletti, P., Bush, A.I. and Sekler, I. (2009). Zinc in the physiology and pathology of the CNS. Nat Rev Neurosci. 10(11): 780-91. https://doi.org/10.1038/nrn2734 PMID: 19826435.

[32] Shuttleworth, C.W. and Weiss, J.H. (2011). Zinc: new clues to diverse roles in brain ischemia. Trends Pharmacol. Sci. 32: 480-486. https://doi.org/10.1016/j.tips.2011.04.001 PMID: 21621864.

[33] Denton, R.M. (2009). Regulation of mitochondrial dehydrogenases by calcium ions. Biochem. Biophys. Acta. 1787: 1309-1326. https://doi.org/10.1016/j.bbabio.2009.01.005 PMID: 19413950.

[34] Takeda, A., Sakurada, N., Ando, M., Kanno, S. and Oku, N. (2009). Facilitation of zinc influx via AMPA/kainate receptor activation in the Hippocampus. Neurochem. Intl. 55: 376-382. https://doi.org/10.1016/j.neuint.2009.04.006.

[35] Stork, C. and Li, Y. (2010). Zinc release from thapsigargin/IP3- sensitive stores in cultured cortical neurons. J. Mol. Signaling. 5: 1-6. https://doi.org/10.1186/1750-2187-5-5.

[36] McCord, M. and Aizenman, E. (2014). The role of intracellular zinc release in aging, oxidative stress, and Alzheimer's disease. Front Aging Neurosci. 6: 77 (1-16). https:// doi.org/10.3389/fnagi.2014.00077.

[37] Jiang, D., Sullivan, P.G., Sensi, S.L., Steward, O. and Weiss, J.H. (2001). Zn(2+) induces permeability transition pore opening and release of pro-apoptotic peptides from neuronal mitochondria. J. Biol. Chem. 276:47524-47529. https://doi.org/ 0.1074/jbc.M108834200.

[38] Dineley, K.E., Malaiyandi, L.M. and Reynolds, I.J. (2002). A reevaluation of neuronal zinc measurements: artifacts associated with high intracellular dye concentration. Mol. Pharmacol. 62: 618-627.https://doi.org/10.1124/mol.62.3.618 PMID: 12181438.

[39] Dineley, K.E., Votyakova, T.V. and Reynolds, I.J. (2003). Zinc inhibition of cellular energy production: implications for mitochondria and neurodegeneration. J. Neurochem. 85: 563-70.https://doi.org/10.1046/j.1471-4159.2003.01678.x PMID: 12694382.

[40] Gazaryan, I.G., Krasinskaya, I.P., Kristal, B.S. and Brown. A.M. (2007). Zinc irreversibly damages major enzymes of energy production and antioxidant defense prior to mitochondrial permeability transition. J. Biol. Chem. 282: 24373-24380. https://doi.org/ DOI 10.1074/jbc.M611376200.

[41] Sindreu C., Storm D.R. (2011). Modulation of neuronal signal transduction and memory formation by synaptic zinc. Front Behav Neurosci. 5: 68. https://doi.org/10.3389/fnbeh.2011.00068 PMID: 22084630.

[42] Quan, X., Thi-Nguyen, T., Choi, S.-K., Xu, S., Das, R., Cha, S.-K., Kim, N., Han, J., Wiederkehr, A., Wollheim, C.B and Park, K-S. (2015). Essential Role of Mitochondrial Ca2+ Uniporter in the Generation of Mitochondrial pH Gradient and Metabolism-Secretion Coupling in Insulin-releasing Cells. J. Biol. Chem. 290, 4086-4096. DOI :https://doi.org/10.1074/jbc.M114.632547.

[43] Oropeza-Almazan, Y. and Blater, L.A. (2020). Mitochondrial calcium uniporter complex activation protects against calcium alternans in atrial myocytes. American Journal of Physiology-Heart and Circulatory Physiology, 319, 4. https://doi.org/10.1152/ajpheart.00375.2020.

[44] Bartolomé, F. and Abramov, A. Y. (2015). Measurement of Mitochondrial NADH and FADAutofluorescence in Live Cells. Metho. Mol. Bio. 1264:263-0. doi.org/10.1007/978-1-4939-2257-4 23.

[45] Colvin, R.A., Fontaine, C.P., Laskowski, M. and Thomas, D. (2003). Zn^{2+} transporters and Zn^{2+} homeostasis in neurons. Eur. J. Pharmacol. 479(1-3): 171-185. https://doi.org/ 10.1016/j.ejphar.2003.08.067 PMID: 14612148.

[46] Li, Y., Hough, C.J., Suh, S.W., Sarvey, J.M. and Frederickson, C.J. (2001a). Induction of mossy fiber, CA3 long-term potentiation requires translocation of synaptically released Zn2+. J. Neurosci. 21: 8015-8025. https://doi.org/270-6474/01/218015-11$15.00/0 PMID: 11588174.

[47] Choi, D.W. and Koh, J.Y. (1998). Zinc and brain injury. Ann. Rev. Neurosci. 21: 347-375. https://doi.org/10.1146/annurev.neuro.21.1.347 PMID: 9530500.

[48] Galasso, S.L. and Dyck, R.H. (2007). The role of zinc in cerebral ischemia. Mol. Med. 13: 380-387.https://doi.org/10.2119/2007-00044.Galasso PMID: 17622314.

[49] Pancani, T., Anderson, K.L., Porter, N.M., and Thibault, O. (2011). Imaging of a glucose analog, calcium and NADH in neurons and astrocytes: dynamic responses to depolarization and sensitivity to pioglitazone. Cell Calcium. 50: 548-558. doi.org/10.1016/j. ceca2011. 09. 002. PMID: 21978418.

[50] Kosterin, P., Kim, G.H., Muschol, M., Obaid, A.L. and Salzberg, B.M. (2005). Changes in FAD and NADH Fluorescence in Neurosecretory Terminals Are Triggered by Calcium Entry and by ADP Production. J. Membr. Biol. 208(2): 113-124.https://doi.org/10.1007/s00232-005-0824-x PMID: 16645741.

Chapter 3

Grid Cells Lose Coherence in Realistic Environments

Yifan Luo, Matteo Toso, Bailu Si, Federico Stella and Alessandro Treves

Abstract

Spatial cognition in naturalistic environments, for freely moving animals, may pose quite different constraints from that studied in artificial laboratory settings. Hippocampal place cells indeed look quite different, but almost nothing is known about entorhinal cortex grid cells, in the wild. Simulating our self-organizing adaptation model of grid cell pattern formation, we consider a virtual rat randomly exploring a virtual burrow, with feedforward connectivity from place to grid units and recurrent connectivity between grid units. The virtual burrow was based on those observed by John B. Calhoun, including several chambers and tunnels. Our results indicate that lateral connectivity between grid units may enhance their "gridness" within a limited strength range, but the overall effect of the irregular geometry is to disable long-range and obstruct short-range order. What appears as a smooth continuous attractor in a flat box, kept rigid by recurrent connections, turns into an incoherent motley of unit clusters, flexible or outright unstable.

Keywords: spatial cognition, realistic environments, grid cells, place cells, recurrent connections

1. Introduction

The nervous system acquires from experience multiple representations of the external world. Extensively studied examples are in the hippocampus and adjoining cortices of rodents and other small mammals, near the apex of their cortical hierarchy [1]. There, the position of the animal in its immediate surroundings and other spatial variables are clearly prominent correlates of neural activity, as exemplified by 50 years of research on place cells [2], but high level representations have been described also for other variables, including time [3, 4], auditory frequency [5], odors [6, 7] and taste [8]. Spatial representations have been thoroughly studied in the laboratory, yielding amazing results [9] but in conditions rather different from those prevailing in the wild. The medial entorhinal cortex (mEC), one synapse upstream of and a major source of inputs to the hippocampus, includes numerous functionally-defined cell types contributing to spatial representations. Significant fractions of its cells have been characterized as grid cells [10], border cells [11], head direction and conjunctive cells [12], speed cells [13] and irregular spatial cells [14]. Should we understand this characterization as encompassing the different components of a precisely engineered system, or as a list of some of the most salient properties expressed by this population of neurons, which is however not rigidly partitioned into cell classes? The question is made more relevant by the failure, over the years, to identify a precise correspondence of such putative classes with cell properties observed in other species, notably in primates.

Place cells fire action potentials when the animal moves through locations in the environment, specific to each cell, called place fields. A simple intuitive model envisages place cells as being assigned, at random, each a location in the environment to represent with its activity, so that at a population level from the list of cells active at a moment in time one can easily decode the position of the animal – who effectively, then, has available a spatial map with its own position annotated on it. This model does not seem outrageously inappropriate, particularly given that the majority of place cells show only one field in the classical laboratory environment, typically smaller than $1m^2$. Several recent experiments, however, show that in larger environments place cells often have multiple and irregularly arranged place fields [15–18]. Still, since the multiple fields are irregularly arranged, one expects that a different list of active cells will uniquely identify each location. Thus place cells, on their own, should effectively represent or map a number of locations in space exponential in the number of cells considered. Even huge environments could be mapped by a sufficient number of place cells.

The effectiveness of the spatial code would appear more doubtful with grid cells, discovered later in the medial entorhinal cortex (mEC) [10]. At least in the simplest intuitive model, each grid cell fires at multiple discrete spaced locations, regularly arranged on a hexagonal pattern that tiles the entire space available to the animal in a laboratory environment. Taken to the extreme, the model would predict that the list of active cells is the same at all locations situated on the hexagonal pattern, which the code would then be unable to distinguish. But are the intuitive models abstracted from experiments in the lab relevant to ecological conditions, where these neural systems have evolved over millions of years?

In this paper, we focus on the pattern formation process of grid cells, based on feed-forward spatial information contributed by place cells, as expressed in our self-organizing adaptation model.

The spatial representations expressed by grid cells and place cells have been reported to differ substantially in the amount of local information they incorporate. Place cells can show global remapping, given sufficient changes in the external environment [19], indicating that they are highly influenced by local spatial information. Grid cells, at least in flat regular environments, do not show global remapping, once the population as a whole has been *anchored* to the local environment [20]. Major changes in the environment, which cause the global remapping of place cells, on an individual basis, appear to induce only a coherent population realignment by grid cells [21]. Subsequent experiments have partially qualified these results, by showing that the grid pattern is influenced by walls [22], unstructured but oddly shaped environments [23], local cues [24], goals [25]. The simple radical notion that grid cells provide a universal spatial metric has therefore been challenged. In addition, in the search of grid patterns in 3D environments, only local order, expressed e.g. by relatively uniform inter-field distances, has been seen to be partially preserved [26], in agreement with the theoretical predictions of the self-organizing adaptation model in 3D [27].

To better understand the spatial selectivity expressed by grid cells, it seems increasingly urgent to move out of artificial laboratory settings. Grid cells have been observed in several species, including rats [10], mice [28] and crawling bats [29], with related cellular selectivity also in monkeys [30] and humans [31, 32], pointing at some degree of universality underlying the phenomenon. These animals experience space in their natural environment, whether large scale 2D where they roam, or 3D where they swim, jump, climb and fly, or curved and crooked, for those who

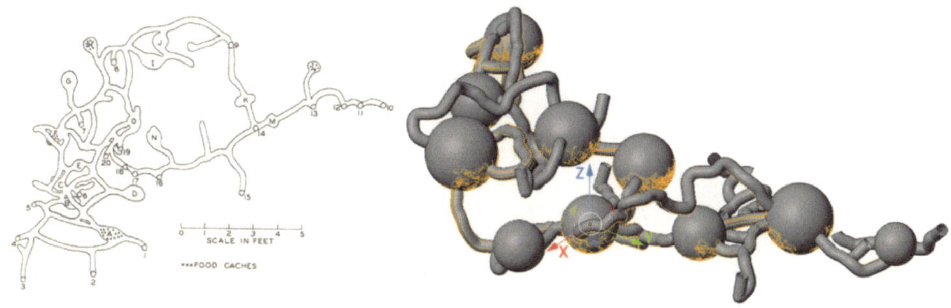

Figure 1.
(Left) Sample map of a Norway rat burrow [40]. (Right) A simulated rat burrow environment which contains 10 sphere representing chambers and 29 tunnels.

live in burrows. One could start by considering three simple types of geometry, flat 2D, 3D, and curved.

Flat 2D [33–36], 3D [27], and curved [37–39] environments have indeed all been considered in studies of the grid pattern emerging with the self-organizing adaptation model.

In this paper, we aim to further our understanding of grid patterns in burrows, where rats, arguably the most frequently used species for the study of spatial representations, usually live in the wild [40, 41].

Norway rats, a most common strain of rats widely used in research, usually referred to as the *common* rat, have seen their burrowing habits meticulously described by John B. Calhoun [40], with the original motivation to control their proliferation in the city of Baltimore. Calhoun has produced estimates of the quantitative characteristics of typical burrows housing ca. 11 adult rats: on average 10 chambers (2 terminal; 8 with at least 2 entrances), linked by 40 tunnels (including on average 13 to exits; 20 internuncial; 7 blind). A sketch is shown in **Figure 1** left [40]. On these descriptions we base our virtual burrows, generated by an *in-house* algorithm, one of which is shown in **Figure 1** right. The yellow line is an example of a simulated trajectory.

2. Simulation

2.1 The construction of the burrows

Our computer model generates virtual burrows with a simple geometry, in which the chambers are represented by spheres, of variable diameter, while the tunnels are schematized as sequences of short curved cylinders (i.e. sections of tori) of variable length and external (curvature) radii, and fixed internal diameter – just enough for a virtual rat to run through. Spheres are thus assigned a center and a diameter, while curved cylinders are defined by a circular basis (with a fixed diameter and a centre and normal versor such that it lies on a sphere, or at the end of another tunnel segment) which is then translated along an arc of circumference (with parameters the versor and magnitude of the curvature radius, and the arc length). Additionally, we draw on the work of Calhoun [40] to define probability distributions for e.g. the diameters of the spheres and the lengths of the tunnels.

Burrow construction proceeds by generating a given number of spheres, with diameters and center-center distances compatible with the observed chamber sizes and distribution of tunnel lengths. Then, the internal burrow connectivity develops, randomly split between chamber-chamber and tunnel-chamber tunnels. Finally, some blind and exit tunnels are added.

The tunnels start with an existing object - sphere or tunnel - and define the initial segment of the new tunnel as a short, straight cylinder with the basis circumference tangent to the preexisting object at a random location, then new curved segments are added, with a basis that coincides with the top of the previous one (same centre, radius and normal versor), while the remaining parameters are randomly generated. The tunnels terminate after randomly growing in random directions for a random length if blind tunnels, when penetrating a target sphere or tunnel if internuncial tunnels, or upon reaching a predefined horizontal plane (i.e. the ground) if exit tunnels. In the last two cases, the direction of growth is biased towards the target destination, with a probability inversely proportional to the distance to the target, and we restrict the choice of the plane of the radial curvature so as to point towards the desired outcome.

2.2 Trajectories

In the simulations, a virtual rat explores one of the virtual burrows described above with a constant speed $v = 40cm/s$. Each time step in the simulation is taken to correspond to $10ms$ in real time. The total length of a simulation is 100 million steps (which would correspond to nearly 12 days of continuous running, to ensure that the self-organization process has approached its asymptote). To obtain smooth random trajectories, resembling those observed in experiments, the change in running direction in the chambers is sampled from a Gaussian distribution with zero mean and standard deviation $\sigma_{RD} = 0.2$ radians; in tunnels, since the size of the tunnels normally can only allow a single rat to pass, the running direction of the virtual rat is always following the tunnels. If the random trajectories lead the virtual rat to the junctions, the virtual rat changes sub-environment (from chamber to tunnel, from tunnel to chamber, or from tunnel to another tunnel). If the tunnel has a dead end, the virtual rat turns back when it reaches the end. The trajectories are limited to the lower half of the environment because of gravity.

2.3 Network model

The model is comprised of two layers. The input layer represents, for example, the CA1 region of the hippocampus and contains $N_{hip} = \rho S + L_t/l_c$ model place cells, which we refer to as *place units* below, where $\rho = 8000/m^2$ is the density of input units in chambers, S is the total area of the chambers, L_t is the total length of the tunnels, l_c is the mean local inter distance of place units in tunnels. This guaranteed that the place units are equally distributed and cover the whole environment. The output layer represents a population of $N_{mEC} = 100$ model mEC cells, which we refer as *grid units* below.

The input to grid unit i at time t is given by

$$h_i^t = \sum_j W_{ij}^t r_j^t \tag{1}$$

The weight W_{ij}^t connects place unit j to grid unit i

Here we assume that grid units develop their maps from scratch, receiving spatially modulated inputs from the place units which have already developed, in line with observations in rat pups, which show that place cells mature earlier than grid cells [42, 43].

Although weak spatial input is sufficient for grid pattern formation [33], regularly arranged place cells are ideal for this function and reduce the averaging necessary for learning with respect to more irregular inputs. Here the activity of each place input unit in space is modeled as a Gaussian place field centered at preferred position $\vec{x}_{j0}$

$$r_j^t = \exp - \left(\frac{\left\| \vec{x}^t - \vec{x}_{j0} \right\|^2}{2\sigma_p^2} \right) \tag{2}$$

where $\vec{x}^t$ is the current location of the virtual rat. $\|\cdot\|$ is the shortest distance (distances in chambers are calculated along great circles, and in tunnels longitudinally, adding them up if $\vec{x}^t$ and $\vec{x}_{j0}$ are not in the same sub-environment). $\sigma_p = 5cm$ is the radius of the place fields.

2.3.1 Single-unit dynamics

The firing rate ψ_i^t of grid unit i is determined through a threshold-nonlinear transfer function

$$\psi_i^t = \psi_{sat} \arctan \left[g^t \left(\alpha_i^t - \mu^t \right) \right] \Theta\left(\alpha_i^t - \mu^t \right) \tag{3}$$

where $\psi_{sat} = 2/\pi$ normalizes the firing rate into arbitrary units. $\Theta(\cdot)$ is the Heaviside function. The variable μ^t is a threshold while α_i^t represents a time-integration of the input h_i, adapted by the dynamical threshold β_i

$$\begin{aligned} \alpha_i^t &= \alpha_i^{t-1} + b_1 \left(h_i^{t-1} - \beta_i^{t-1} - \alpha_i^{t-1} \right), \\ \beta_i^t &= \beta_i^{t-1} + b_2 \left(h_i^{t-1} - \beta_i^{t-1} \right) \end{aligned} \tag{4}$$

where β_i has slower dynamics than α_i, and b_2 is set to $b_2 = b_1/3$, $b_1 = 0.1$. These adaptive dynamics make it more difficult for a neuron to fire for a long period of time, and endow grid units with fatigue dynamics [33]. The gain g^t and threshold μ^t are iteratively adjusted at every time step to fix the mean activity $a = \sum_i \psi_i^t / N$ and the sparsity $s = \left(\sum_i \psi_i^t \right)^2 / \left(N \sum_i \psi_i^{t2} \right)$ within a 10% relative error bound from pre-specified values, $a_0 = 0.1$ and $s_0 = 0.3$ respectively.

2.3.2 Head direction modulation and collateral connections

Head direction (HD) modulation and collateral connections are important for grid alignment, as suggested by the detailed analysis in [33, 34]. The head direction in the chambers (spheres) is defined as the angle between a vector and the vector pointing towards the north pole.

With the addition of HD modulation and collateral connections, Eq. (1) for the inputs to grid unit i is rewritten

$$h_i^t = f_{\theta_i}(\omega_t) \left(\sum_j W_{ij}^{t-1} r_j^t + \rho_1^t \sum_k W_{ik}^{t-1} \psi_k^{t-\tau} \right) \tag{5}$$

where $\psi_k^{t-\tau}$ is the activity of other grid unit k reverberated by collateral connections W_{ik}^t with a delay $\tau = 25$ steps. $\rho_1^t = \varphi t/T$ when $t < T$ and φ when $t \geq T$, where $T = 5 \times 10^6$ for each simulation, and φ is a set value controlling the strength of recurrent connections. Then the time-dependent strength ρ_1^t is gradually increasing from zero, in order to reduce the influence of the initial random weights.

$f_{\theta_i}(\omega_t)$ is the HD tuning function that has maximal value when the current head direction ω_t of the simulated rat is along the preferred direction θ_i.

$$f_\theta(\omega) = c + (1 - c)\exp\left(v(\cos(\theta - \omega) - 1)\right) \tag{6}$$

and $c = 0.1$ and $v = 0.8$ are parameters determining the baseline activity and the width of head direction tuning.

2.3.3 Synaptic plasticity

All weights in the network self-organize while the virtual rat explores the environment and the updating following the Hebbian rule.

Weights between place units and grid units are changed according to

$$\Delta W_{ij}^t = \varepsilon^t \left(\psi_i^t r_j^t - \overline{\psi}_i^{t-1} \overline{r}_j^{t-1}\right) \tag{7}$$

where $\varepsilon^t = \xi(1 - 0.9t/T)$ when $t < T$ and 0.1ξ when $t \geq T$, here $T = 5 \times 10^6$, $\xi = 0.01$. $\overline{\psi}_i^t$ and $\overline{r}_j^t$ are estimated mean firing rates.

$$\begin{aligned} \overline{\psi}_i^t &= \overline{\psi}_i^{t-1} + \eta\left(\psi_i^t - \overline{\psi}_i^{t-1}\right), \\ \overline{r}_i^t &= \overline{r}_i^{t-1} + \eta\left(r_i^t - \overline{r}_i^{t-1}\right), \end{aligned} \tag{8}$$

and $\eta = 0.05$ is a positive averaging factor.

The collateral weights between grid units are adapted according to

$$\Delta W_{ik}^t = \zeta \psi_i^t \left(\psi_i^{t-\tau} - \kappa\right) \tag{9}$$

Here $\zeta = 6.6 \times 10^{-5}$ is a learning rate smaller, at least initially, than the learning rate for feed-forward weights between place units and grid units. $\kappa = 0.1$ is an inhibition factor.

All the weights in the network are initialized as random numbers $(1 - \gamma) + \gamma u$. $\gamma = 0.1$, and u is a random variable uniformly distributed in $[0, 1]$.

After initialization or weight changes, all weights are normalized to a unitary L_2 norm

$$\sum_j {W_{ij}^t}^2 = 1 \tag{10}$$

3. Results

In the simulations, grid units may have been expected to form highly regular patterns, as they do in flat 2D environments [33]. However, in our burrows, modeling the natural environment of real rats, the activity patterns that get established are not nearly as regular. **Figure 2** shows the grid map of one sample grid unit. The lower half of the sphere, representing a chamber, was projected to a horizontal plane, while the upper half was discarded, as trajectories are limited to the lower half to model gravity. Tunnels were straightened into 1D segments. Sufficiently explored chambers show clearly identifiable fields, while those where the virtual rat spent less time show only blurred, often overlapping fields. Longer

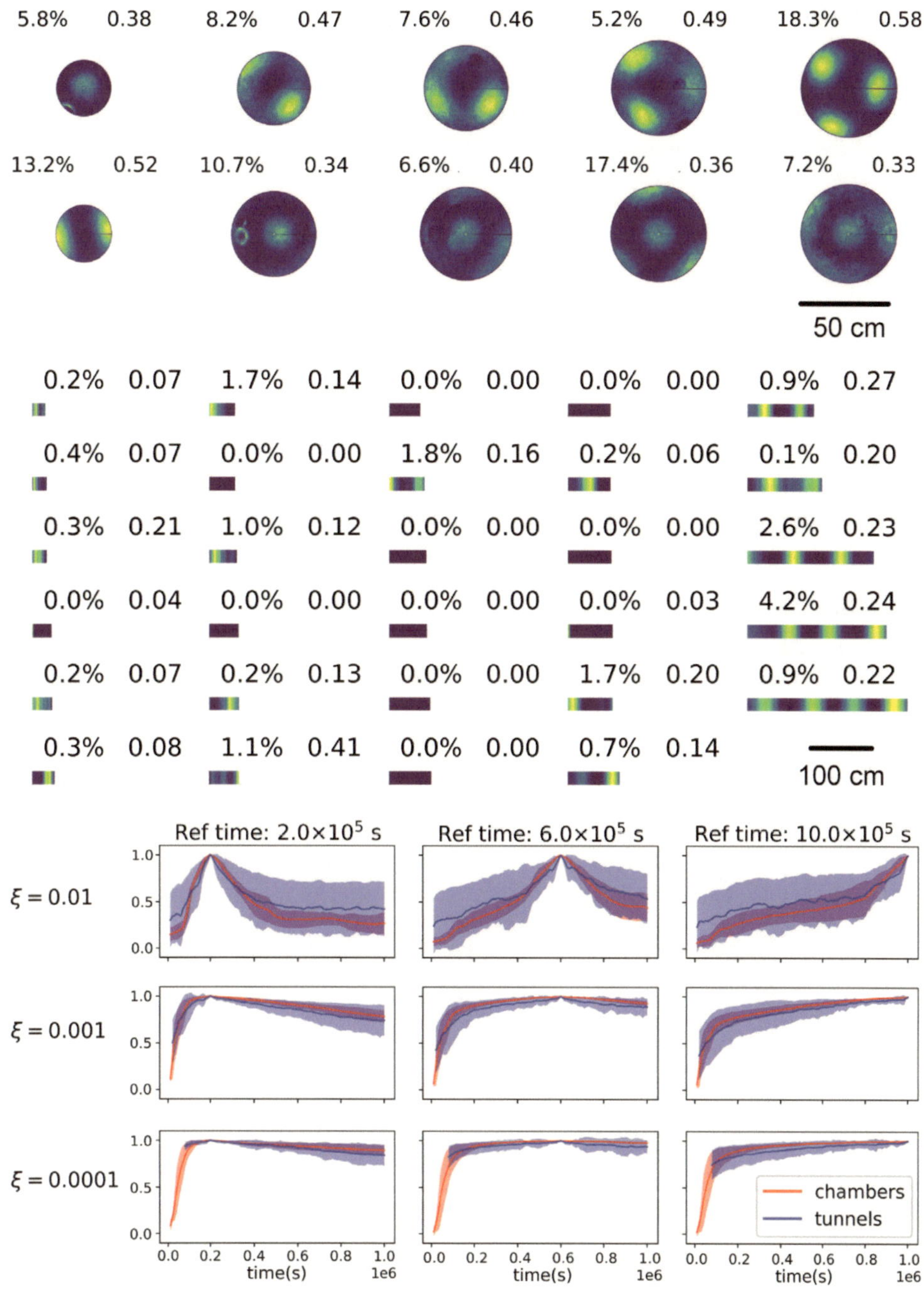

Figure 2.
Grid maps in burrows and their stability. (Top and center) The maps of one unit in chambers (the lower half of each chamber is shown), and in tunnels (straightened and taken to be 1D). Chambers (tunnels) are shown in order of radius (length), with the percent time spent in each by the virtual rat indicated top left, top right indicates the maximum firing rate (always in the range [0,1], given Eq. (3)). (Bottom) Correlation with the map at the reference time. From the first to the third row, the initial learning rate is lowered from $\xi = 0.01$ through $\xi = 0.001$ to $\xi = 0.0001$; while from the first to the third column, the reference time points are 2×10^5, 6×10^5, 10×10^5.

simulations make for little improvement. Tunnels, as they are 1D structures, are easier to learn and show clear fields even in shorter simulations.

The grid maps in a chamber include very few fields, as indicated in **Figure 2** top. In fact, chamber width is at most 0.3 m (reported as 298 mm, with median 221 mm and minimum 155 mm [40]), unlike the 2 m diameter of the flat circular enclosure used in the Moser lab [10], and even considerably smaller than the small square boxes used earlier and in several later studies (e.g., $1.0 \times 1.0m$ [44]). Due to this reason, to be able to study the layout of the fields in a spherical chamber we have used larger diameters in our model curved environments, so they would include e.g. the 12 fields of the most regular pentagonal arrangement [37, 39] (on the entire sphere, top and bottom, with 6 in the lower half).

For grid patterns to be stable, the feed-forward learning rate between place units and grid units has to be very small. As shown in **Figure 2** bottom, with learning rate $\xi = 0.01$ grid units may keep shifting their fields as the simulation proceeds, so that their correlation with those at any reference time keeps changing; with learning rate $\xi = 0.001$ the stability is markedly improved, and with $\xi = 0.0001$ grid units appear to form stable maps both in chambers and in tunnels.

The gridness score has been widely used to quantify the spatial periodicity of grid patterns, but it can be applied only to patterns with six-fold symmetry. Results from simulations in curved environments show regular grid patterns with five-fold or lower symmetry for constant positive curvature, and seven-fold or higher symmetry for negative curvature [37, 38]. In the burrows simulated here, inter-field distances vary also within chambers (light blue distribution in **Figure 3** top), indicating that spatial periodicity is not a property of grid cells in natural environments.

The inter field distances in tunnels (green distribution in **Figure 3** top) have a larger peak value than in chambers. In chambers, in fact, our simulated trajectories are curved as the running direction keeps changing, unlike the trajectories in tunnels which can only follow the 1D tunnels, so the distance traveled over the adaptation time scale is longer in tunnels than in chambers. For real rats, of course, the distance traveled in each sub environment depends strongly on their prevailing

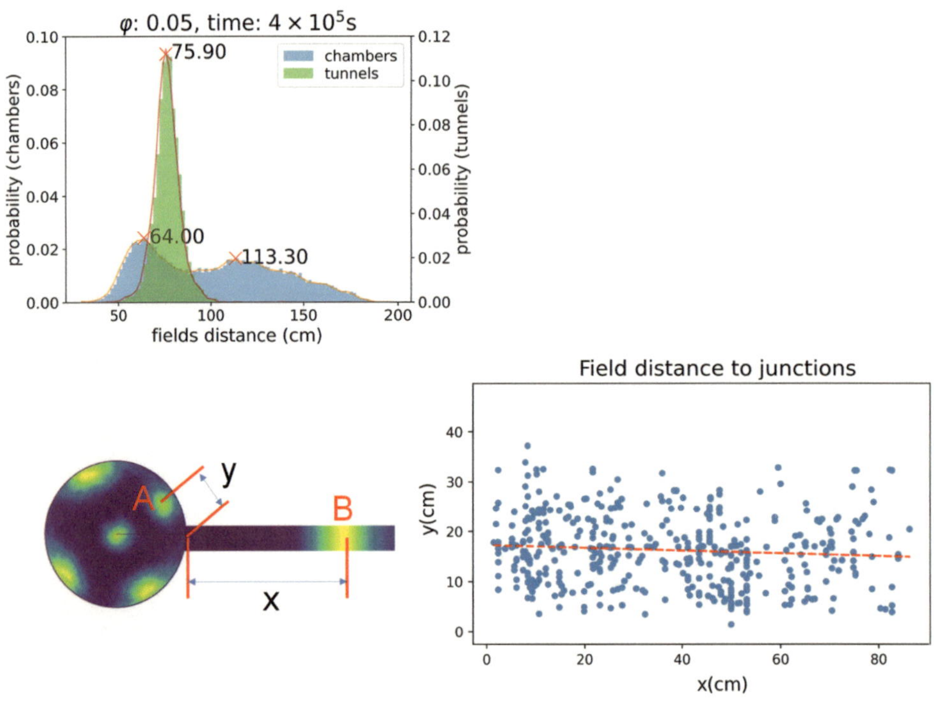

Figure 3.
(Top) Inter fields distributions of fields in the same chambers and tunnels, indicating the location of the peaks. (Bottom left) Field A and field B are the closest fields to a junction, and y and x represent their distance to the junction from the chamber and tunnel side. (Bottom right) The correlation between the x and y measures.

speed, likely contributing to the different representation of grid cells in tunnels and chambers.

The virtual rat learns the entire environment at the same time, since the randomly generated trajectories span it all. We asked, then, whether grid units form a continuous representation of the whole environment, by checking whether the junctions connecting chambers and tunnels break the continuity. In **Figure 3** bottom left, field B and field A are the closest fields to a junction, from the tunnel and chamber side, respectively, and x and y represent their distance from the junction, so that $x =$ field distance $- y$. As shown in **Figure 3** bottom right, x and y do not show the expected negative correlation (with -1 or with any other clear slope), but rather a loose relationship. This indicates that spatial representations by simulated grid units are effectively independent in distinct portions of the environment, as this is in practice partitioned up by the junctions.

Realistic simulated burrows like that in **Figure 1** right require enormous CPU time to be explored with sufficient statistics, and even then the maps that form especially in the chambers often present rather unclear fields, as in **Figure 2**. Therefore, in the following analyses we consider a *simplified* burrow, with only three chambers and three tunnels connecting them, and compare the maps emerging there with those in a square box or in/on a sphere. Since the gridness score cannot be readily applied, we use instead the distributions of inter field distance (left column of **Figure 4**) and of the angles between triplets of nearby fields (right column of **Figure 4**); such triplets are defined by mutual distance in the range of 50–150% of the first peak in the field distance distribution. We generated data for the square box environment setting the strength of recurrent connections at $\varphi = 0.125$, much stronger than in the sphere and burrow, where it was set at $\varphi = 0.05$, in order to have in each environment as regular grid patterns as they could be (see **Figure 5**). The effect of recurrent strength in different environments will be described in details later. Other parameters were kept constant.

The inter-field distance distribution in the box environment (of size $1.5 \times 1.5 m^2$) has clear peaks, shown in **Figure 4** upper left. The fourth and fifth peaks are shifted to the left compared with those of perfectly regular six-fold grids, probably because the virtual trajectories are limited by the hard border (which reflects the trajectory as in a mirror when it hits the border). The angles have a distribution centered at 60 degrees, shown in **Figure 4** upper right.

The spherical environment, with radius $r = 0.5m$, allows most grid units to develop 12 fields. Their distribution of field distances also has clear peaks (**Figure 4** center left). The angle distribution is centered at 71.8 degrees (**Figure 4** center right), just 0.2 degrees below the 72 degrees value of the perfectly regular five-fold grid pattern.

In the burrow environment, simulations have parameters consistent with those in the sphere. The field distance distribution does not have clear peaks, however (**Figure 4** lower left). The angle distribution shows more variability than in the box and in the sphere (**Figure 4** lower right). This suggests that the formation of regular grid patterns in natural environments, like rat burrows, is very challenging for the same system that produces them easily in laboratory conditions.

The grid maps of cells recorded at the same electrode position show generally a small correlation or rather an anti-correlation [10], because even with similar grid spacing and orientation, a relative phase shift between cells is sufficient to remove the correlation between them. The results from simulations in the flat box environment point at the same phenomenon, with in fact most pairs of grid units ending up negatively correlated, as shown in **Figure 6** top. Away from six-fold symmetry, however, things are a bit different. In the sphere, units can form five-fold symmetric grid patterns, when they have 12 fields, fields which *cannot* be translated on top

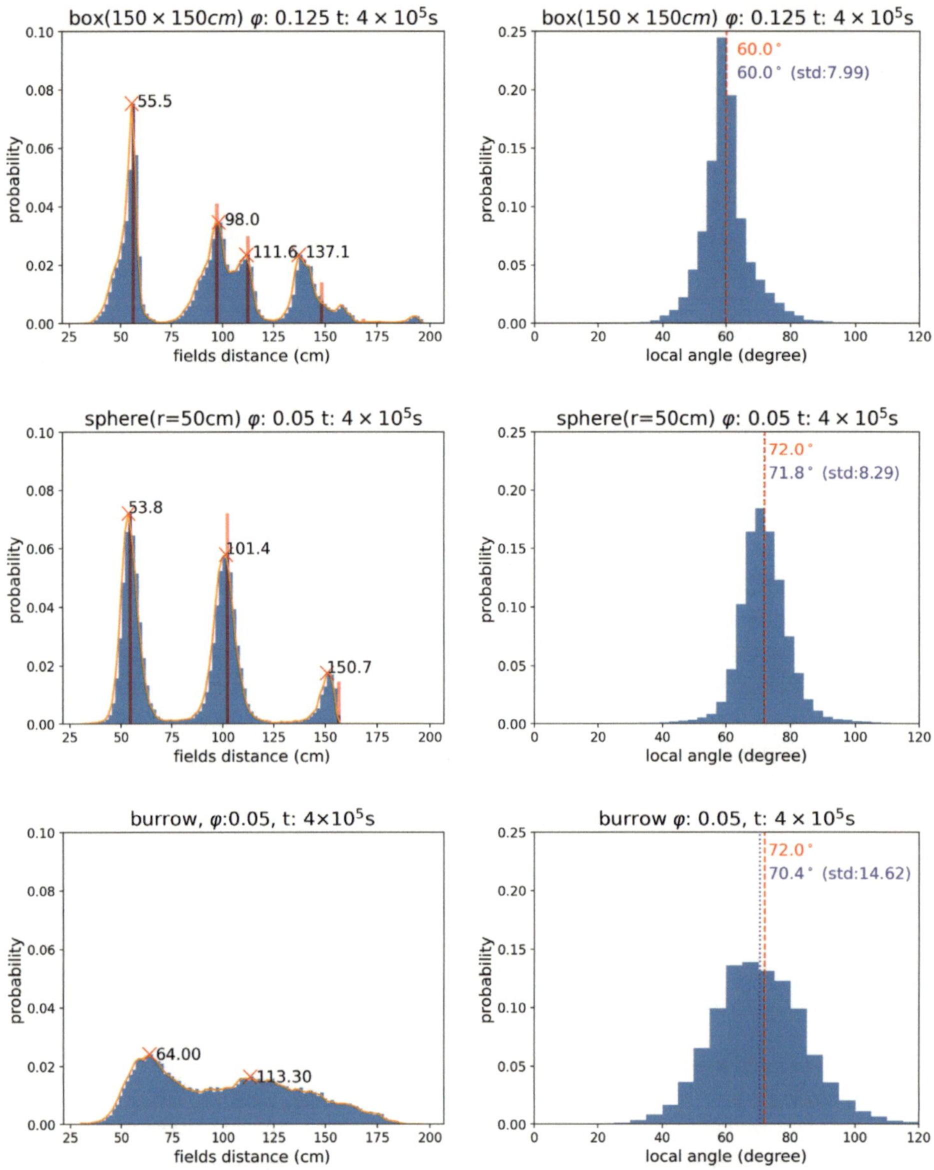

Figure 4.
Inter field distance and angles from local triangles. (Left) Distribution of inter field distance in box, spherical and burrow environments. Shown in red for the box and sphere is the distribution valid for regular grids, with the first peak aligned with that in our simulations. All peaks are indicated. (Right) The angle distribution, where again the red dotted curve refers to regular grids; the blue indicates simulation data, with standard deviation are annotated.

of each other, because translations do not exist on curved surfaces. So the bulk of the units are less anti-correlated, and the peak correlation shifts to less negative. In the burrow, without regular grid patterns, map correlations show a peak located at even higher values, near zero.

Both in the box and sphere environments, the correlation between *some* units reaches to almost 1, while most pairs of units are actually anti-correlated. The distribution is much less spread out in the burrow environment where, as we said (**Figure 6** top), most pairs of units have correlation close to zero, and the most correlated ones hardly reach above 0.6 spatial (Pearson) correlation. Despite this, the examples in **Figure 6** bottom show that a standard K-means algorithm identifies

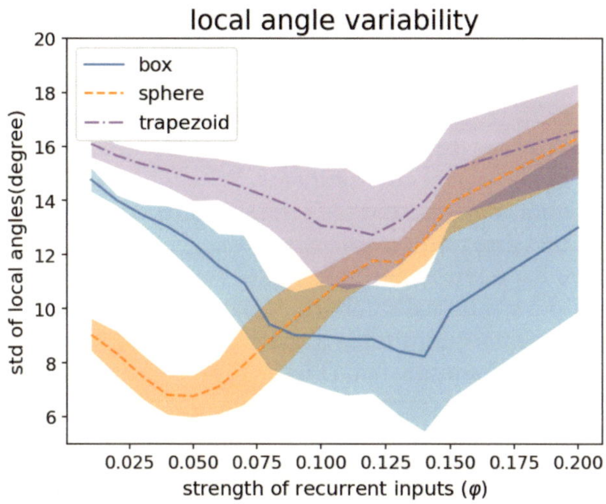

Figure 5.
Standard deviation in local angles for the square box, trapezoid, and sphere environments varying the strength of recurrent connections. The square box environment is 150cm × 150cm, while the trapezoid has parallel walls 87cm and 174cm long, with symmetric 179.4cm side walls. The sphere has a 50cm radius, suited to include 12 fields per unit. The solid line is the mean value of the standard deviation, the shaded regions indicate its own standard deviation.

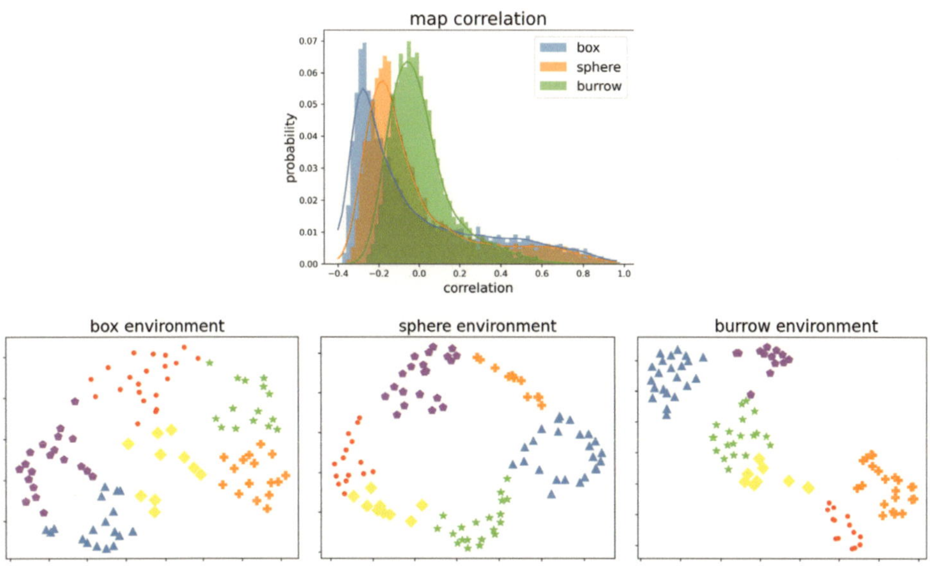

Figure 6.
(Top) Map correlation between pairs of simulated units, in different environments. What is shown is the distribution of Pearson correlation values across pairs. (Bottom) Map clusters visualized with t-distributed stochastic neighbor embedding method (TSNE). The map distance between pair of units is defined by 1 minus their Pearson correlation, and units were clustered with a K-means algorithm into 6 clusters.

6 clusters of units in the burrow data, with the same or even greater ease than in the box or sphere data. It appears that the reason is that in the sphere, and even more in the box, different units map out with their fields a low-dimensional spatial continuum, so that breaking them into clusters is quite arbitrary – an intuitive example would be that of grid units with fields equi-spaced on a 1D ring, that would themselves describe a ring if randomly spread out, and not a clustered structure. In the

multi-chamber environment of the simplified burrow, clusters apparently emerge spontaneously (see **Figure 6** bottom).

Local recurrent connections have been identified as a key element of grid pattern formation [45], in particular to align the grids in a flat environment [35, 46, 47]. We asked how the strength of recurrent connections affects the pattern formation process in our model environments. We simulated the adaptation model in flat and curved environments, and took the standard deviation of the angles from local triangles as a measure of the regularity of the grid pattern.

Considering two flat environments, a square box and a trapezoid with the same area, grid patterns show minimal standard deviation in the square box when the strength of recurrent connections takes a specific value $\varphi = 0.14$, as shown in **Figure 5**; and a larger minimum value (**Figure 5**) for somewhat weaker recurrent connections ($\varphi = 0.12$) in the trapezoid, known to distort the grid pattern of real cells [23].

In the sphere, however, the minimum standard deviation, comparable with that in the box environment (as both allow for regular tessellation, five-fold in one case and six-fold in the other), is reached with much weaker recurrent connections, $\varphi = 0.05$, as shown in **Figure 5**. In even the simplified burrow, the standard deviation of the angle distribution is inherently higher, because of the lack of long-range order even in the absence of recurrent connections, and recurrent connections are bound to increase the irregularity further (see **Figure 4** lower left).

In general, the strength of the recurrent connections might contribute to the rigidity of the grid pattern across environments, but also perhaps to its flexibility in the interaction with walls [22], changes in the boundary [48] and the appearance of local cues [24], including goals [25].

4. Discussion

The structure of the natural habitat of any species would appear to be a prime determinant of exactly how that species has adapted to live in that habitat; yet in exploring the spatial memory and navigation abilities in rodents, and in particular the neural systems that subserve them, early experiments have focused on artificial laboratory environments, incongruent with those prevailing in the wild. The discovery of the remarkable spatial selectivity of grid cells, evident in such laboratory set-ups, has raised the issue of what grid cell firing patterns would look like, in an ecological setting.

More recent experiments have pointed at increased irregularity and plasticity of grid activity patterns, whether due to large environments [20], non-standard shapes [23], modulation by local cues [24], boundary changes [48] or the presence of goals [25]. These observations, however, were largely framed as deviations or perturbations from the ideal notion of a regular tessellation of the environment, exhibiting long-range order *ad infinitum*, which had been evoked by the early findings.

The possibility that long-range order may not apply at all in an ecological setting was raised initially by looking at the activation patterns that emerge, with the adaptation model, in 3D [27] or on curved 2D surfaces [37, 38], and is confirmed in a long-running experimental study in bats [26]. While bats fly, rats are burrowing animals, and natural burrows are much more complicate structures than open arenas or other schematic laboratory settings.

In the present study, we have modeled the burrow environment based on the detailed quantitative descriptions by John B. Calhoun, and we have let a virtual rat randomly explore it; then using our self-organizing adaptation model we have observed grid pattern formation in model units. Grid units can attain stable representation of the whole environment, if acquiring it slowly, but less explored sub environments tend to be represented by blurred maps. The limited size of the chambers in natural burrows only allows grid units to express very few fields, challenging the very idea that grid cells may show long-range order outside the lab. One may wonder whether the reports of a six-fold symmetry in imaging data from humans, a putative signature of an underlying grid-like representation, might be due to large virtual arenas used in those studies [31]. It comes as no surprise, then, that the characteristic signature was not observed in the limited and non-flat vowel space [49], although of course there may be many other reasons for a null result.

The continuity of spatial representations depends on the exploration. In [50], with a multi-compartment environment, real grid cells firing patterns could establish a single, continuous representation that spanned both compartments after prolonged experience. In natural burrows however, tunnels are so narrow that they are effectively 1D structures, and as such they necessarily break any potential continuity in the representation of the chambers. What is left, at least when studied with the adaptation model, is effectively a representation in terms of disjoint spatial *fragments*, which coincide, in our modeling framework, with the better explored chambers.

The fragmentary nature of these spatial representations finds expression also at the level of neural populations. Compared with simulations in regular environments (in the square box, but also to some extent in the large hemi-sphere which can accommodate 6 fields), grid units simulated in burrows have more of a tendency to cluster in groups with similar fields. This might be part of the drive that leads to the observed *modularity* of grid activity [36, 51].

In flat environments recurrent connections promote regularity (and can align out-of-spatial-phase patterns). In simple environments with constant non-zero curvature, recurrent connections promote irregularity [39]. In real-life environments, it is likely that they cause both irregularity and clustering into groups of units with similar selectivity, running against the principle of representing space evenly, expressed in the most idealized conceptual model by the notion of a continuous attractor.

Experiments in artificial laboratory settings, which utilize well-controlled and simplified paradigms, thus gave us the opportunity to admire a most impressive feat of the rodent nervous system, its ability to approximate a regular tiling of an infinite plane. While an exhilarating experience, this may have distracted us from understanding the characteristics of grid cells in ecological conditions, and possibly their function in an evolutionary perspective.

Author details

Yifan Luo[1], Matteo Toso[2], Bailu Si[3], Federico Stella[4] and Alessandro Treves[1]*

1 Sector of Cognitive Neuroscience, SISSA, Trieste, Italy

2 PAVIS, Istituto Italiano di Tecnologia, Genova, Italy

3 School of Systems Science, Beijing Normal University, Beijing, China

4 Donders Institute for Behaviour and Cognition, Radboud University, Nijmegen, Netherlands

*Address all correspondence to: ale@sissa.it

References

[1] Felleman DJ, Van Essen DC. Distributed Hierarchical Processing in the Primate Cerebral Cortex. Cerebral Cortex. 1991 Jan;1(1):1–47.

[2] O'Keefe J, Dostrovsky J. The Hippocampus as a Spatial Map. Preliminary Evidence from Unit Activity in the Freely-Moving Rat. Brain Res. 1971 Nov;34(1):171–175.

[3] MacDonald CJ, Lepage KQ, Eden UT, Eichenbaum H. Hippocampal "Time Cells" Bridge the Gap in Memory for Discontiguous Events. Neuron. 2011 Aug;71(4):737–749.

[4] Buzsáki G, Tingley D. Space and Time: The Hippocampus as a Sequence Generator. Trends in Cognitive Sciences. 2018 Oct;22(10):853–869.

[5] Aronov D, Nevers R, Tank DW. Mapping of a Non-Spatial Dimension by the Hippocampal– Entorhinal Circuit. Nature. 2017 Mar;543(7647):719–722.

[6] Eichenbaum H, Kuperstein M, Fagan A, Nagode J. Cue-Sampling and Goal-Approach Correlates of Hippocampal Unit Activity in Rats Performing an Odor-Discrimination Task. Journal of Neuroscience. 1987 Mar;7(3):716–732.

[7] Taxidis J, Pnevmatikakis EA, Dorian CC, Mylavarapu AL, Arora JS, Samadian KD, et al. Differential Emergence and Stability of Sensory and Temporal Representations in Context-Specific Hippocampal Sequences. Neuron. 2020 Dec;108(5):984–998.e9.

[8] Herzog LE, Pascual LM, Scott SJ, Mathieson ER, Katz DB, Jadhav SP. Interaction of Taste and Place Coding in the Hippocampus. Journal of Neuroscience. 2019 Apr;39(16): 3057–3069.

[9] Moser EI, Moser MB, McNaughton BL. Spatial Representation in the Hippocampal Formation: A History. Nature Neuroscience. 2017 Oct; 20(11):1448–1464.

[10] Hafting T, Fyhn M, Molden S, Moser MB, Moser EI. Microstructure of a Spatial Map in the Entorhinal Cortex. Nature. 2005 Aug;436(7052):801–806.

[11] Solstad T, Boccara CN, Kropff E, Moser MB, Moser EI. Representation of Geometric Borders in the Entorhinal Cortex. Science. 2008 Dec;322(5909): 1865–1868.

[12] Sargolini F, Fyhn M, Hafting T, McNaughton BL, Witter MP, Moser MB, et al. Conjunctive Representation of Position, Direction, and Velocity in Entorhinal Cortex. Science. 2006 May;312(5774):758–762.

[13] Kropff E, Carmichael JE, Moser MB, Moser EI. Speed Cells in the Medial Entorhinal Cortex. Nature. 2015 Jul;523 (7561):419–424.

[14] Diehl GW, Hon OJ, Leutgeb S, Leutgeb JK. Grid and Nongrid Cells in Medial Entorhinal Cortex Represent Spatial Location and Environmental Features with Complementary Coding Schemes. Neuron. 2017 Apr;94(1):83–92.e6.

[15] Fenton AA, Kao HY, Neymotin SA, Olypher A, Vayntrub Y, Lytton WW, et al. Unmasking the CA1 Ensemble Place Code by Exposures to Small and Large Environments: More Place Cells and Multiple, Irregularly Arranged, and Expanded Place Fields in the Larger Space. The Journal of Neuroscience. 2008 Oct;28(44):11250–11262.

[16] Park E, Dvorak D, Fenton AA. Ensemble Place Codes in Hippocampus: CA1, CA3, and Dentate Gyrus Place Cells Have Multiple Place Fields in Large Environments. PLOS ONE. 2011 Jul;6(7):e22349.

[17] Rich PD, Liaw HP, Lee AK. Large Environments Reveal the Statistical Structure Governing Hippocampal Representations. Science. 2014 Aug;345 (6198):814–817.

[18] Harland B, Contreras M, Souder M, Fellous JM. Dorsal CA1 Hippocampal Place Cells Form a Multi-Scale Representation of Megaspace. Current Biology. 2021 May;31(10):2178–2190.e6.

[19] Latuske P, Kornienko O, Kohler L, Allen K. Hippocampal Remapping and Its Entorhinal Origin. Frontiers in Behavioral Neuroscience. 2018;11.

[20] Stensola T, Stensola H, Moser MB, Moser EI. Shearing-Induced Asymmetry in Entorhinal Grid Cells. Nature. 2015 Feb;518(7538):207–212.

[21] Fyhn M, Hafting T, Treves A, Moser MB, Moser EI. Hippocampal Remapping and Grid Realignment in Entorhinal Cortex. Nature. 2007 Mar; 446(7132):190–194.

[22] Barry C, Hayman R, Burgess N, Jeffery KJ. Experience-Dependent Rescaling of Entorhinal Grids. Nature Neuroscience. 2007 Jun;10(6):682–684.

[23] Krupic J, Bauza M, Burton S, Barry C, O'Keefe J. Grid Cell Symmetry Is Shaped by Environmental Geometry. Nature. 2015 Feb;518(7538):232–235.

[24] Ismakov R, Barak O, Jeffery K, Derdikman D. Grid Cells Encode Local Positional Information. Current Biology. 2017 Aug;27(15):2337â32343.e3.

[25] Boccara CN, Nardin M, Stella F, O'Neill J, Csicsvari J. The Entorhinal Cognitive Map Is Attracted to Goals. Science. 2019 Mar;363(6434):1443–1447.

[26] Ginosar G, Aljadeff J, Burak Y, Sompolinsky H, Las L, Ulanovsky N. Locally Ordered Representation of 3D Space in the Entorhinal Cortex. Nature. 2021 Aug:1–6.

[27] Stella F, Treves A. The Self-Organization of Grid Cells in 3D. eLife. 2015 Mar;4:e05913.

[28] Fyhn M, Hafting T, Witter MP, Moser EI, Moser MB. Grid Cells in Mice. Hippocampus. 2008 Dec;18(12): 1230–1238.

[29] Yartsev MM, Witter MP, Ulanovsky N. Grid Cells without Theta Oscillations in the Entorhinal Cortex of Bats. Nature. 2011 Nov;479 (7371):103–107.

[30] Killian NJ, Jutras MJ, Buffalo EA. A Map of Visual Space in the Primate Entorhinal Cortex. Nature. 2012 Nov; 491(7426):761–764.

[31] Doeller CF, Barry C, Burgess N. Evidence for Grid Cells in a Human Memory Network. Nature. 2010 Feb; 463(7281):657–661.

[32] Jacobs J, Weidemann CT, Miller JF, Solway A, Burke JF, Wei XX, et al. Direct Recordings of Grid-like Neuronal Activity in Human Spatial Navigation. Nature Neuroscience. 2013 Sep;16(9): 1188–1190.

[33] Kropff E, Treves A. The Emergence of Grid Cells: Intelligent Design or Just Adaptation? Hippocampus. 2008 Dec;18 (12):1256–1269.

[34] Si B, Kropff E, Treves A. Grid Alignment in Entorhinal Cortex. Biological Cybernetics. 2012 Aug;106(8–9):483–506.

[35] Si B, Treves A. A Model for the Differentiation between Grid and Conjunctive Units in Medial Entorhinal Cortex. Hippocampus. 2013 Dec;23(12): 1410–1424.

[36] Urdapilleta E, Si B, Treves A. Self-Organization of Modular Activity of Grid Cells. Hippocampus. 2017 Aug.

[37] Stella F, Si B, Kropff E, Treves A. Grid Cells on the Ball. Journal of Statistical Mechanics: Theory and Experiment. 2013;2013(03):P03013.

[38] Urdapilleta E, Troiani F, Stella F, Treves A. Can Rodents Conceive Hyperbolic Spaces? Journal of The Royal Society Interface. 2015 Jun;12(107): 20141214.

[39] Stella F, Urdapilleta E, Luo Y, Treves A. Partial Coherence and Frustration in Self-Organizing Spherical Grids. Hippocampus. 2020;30(4): 302–313.

[40] Calhoun JB. The Ecology and sociology of the Norway Rat. U. S. Government Printing; 1962.

[41] Schweinfurth MK. The Social Life of Norway Rats (Rattus Norvegicus). eLife. 2020 Apr;9:e54020.

[42] Langston RF, Ainge JA, Couey JJ, Canto CB, Bjerknes TL, Witter MP, et al. Development of the Spatial Representation System in the Rat. Science. 2010 Jun;328(5985): 1576–1580.

[43] Wills T, Cacucci F, Burgess N, O'Keefe J. Development of the Hippocampal Cognitive Map in Pre-Weanling Rats. Science (New York, NY). 2010 Jun;328(5985):1573–1576.

[44] Fyhn M, Molden S, Witter MP, Moser EI, Moser MB. Spatial Representation in the Entorhinal Cortex. Science. 2004 Aug;305(5688): 1258–1264.

[45] Couey JJ, Witoelar A, Zhang SJ, Zheng K, Ye J, Dunn B, et al. Recurrent Inhibitory Circuitry as a Mechanism for Grid Formation. Nature Neuroscience. 2013 Mar;16(3):318–324.

[46] D'Albis T, Kempter R. Recurrent Amplification of Grid-Cell Activity. Hippocampus. 2020;30(12):1268–1297.

[47] Tukker JJ, Beed P, Brecht M, Kempter R, Moser EI, Schmitz D. Microcircuits for Spatial Coding in the Medial Entorhinal Cortex. Physiological Reviews. 2021 Jul.

[48] Wernle T, Waaga T, Mørreaunet M, Treves A, Moser MB, Moser EI. Integration of Grid Maps in Merged Environments. Nature Neuroscience. 2017 Dec:1.

[49] Kaya Z, Soltanipour M, Treves A. Non-Hexagonal Neural Dynamics in Vowel Space. AIMS Neuroscience. 2020; 7(neurosci-07-03-015):275–298.

[50] Carpenter F, Manson D, Jeffery K, Burgess N, Barry C. Grid Cells Form a Global Representation of Connected Environments. Current Biology. 2015 May;25(9):1176–1182.

[51] Stensola H, Stensola T, Solstad T, Froland K, Moser MB, Moser EI. The Entorhinal Grid Map Is Discretized. Nature. 2012 Dec;492(7427):72–78.

Chapter 4

Hippocampal Influences on Movements, Sensory, and Language Processing: A Role in Cognitive Control?

Douglas D. Burman

Abstract

Beyond its established role in declarative memory function, the hippocampus has been implicated in varied roles in sensory processing and cognition, particularly those requiring temporal or spatial context. Disentangling its known role in memory from other cognitive functions can be challenging, as memory is directly or indirectly involved in most conscious activities, including tasks that underlie most experimental investigations. Recent work from this lab has examined the directional influence from the hippocampus on cortical areas involved in task performance, including tasks requiring movements, sensory processing, or language judgments. The hippocampus shows preferential connectivity with relevant cortical areas, typically the region critically involved in task performance, raising the possibility that the hippocampus plays a role in cognitive control. Minimal criteria for a role in cognitive control are proposed, and hippocampal connectivity with sensorimotor cortex during a non-mnemonic motor task is shown to meet this standard. Future directions for exploration are discussed.

Keywords: hippocampus, cortex, connectivity, PPI, cognitive control, sensorimotor, language

1. Introduction

Since its earliest description in 1587, many different functions have been ascribed to the hippocampus, each based on the available techniques and prevailing understanding of brain function. The earliest hypotheses were based on its observed anatomical connections. The hippocampus was first believed to be olfactory, based on erroneous observations suggesting direct olfactory input [1]. Olfactory input to the hippocampus is in fact indirect; except for a role in odorous memories, olfaction is no longer believed to be the hippocampus' primary function.

By the early twentieth century, a role of the brain in emotional and cognitive states had been well-established, and procedures were developed to better trace brain pathways and identify brain lesions. The hippocampus was identified as one structure within the "limbic lobe". Including the entire hippocampal formation, cingulate gyrus, and associated areas, Papez theorized this region to be involved in the expression of emotional behaviors [2, 3]. Support for this idea was seen in the experiments of Klüver and Bucy, who reported that resection of the medial temporal lobe (including the hippocampal formation and nearby amygdaloid complex) had extreme effects on emotional behaviors [4–6].

In the 1950s, the spontaneous activity of the hippocampus was noted to bear a consistent relationship to various states of consciousness [7], generating several hypotheses about high-level cognitive functions. These ideas were largely dismissed, as researchers had demonstrated that lower mammals could still function (albeit with deficits) after the hippocampus was experimentally removed [8]. Anatomical studies further refined our knowledge of hippocampal connections across the brain. After several stages of processing, information from every sensory modality funnel into the hippocampus via the entorhinal cortex, with multiple senses sometimes combined; the hippocampus indirectly projects to the entire cerebral cortex, mostly via the subiculum [9].

When Scoville and Milner surgically resected a patient's hippocampus in an attempt to relieve epileptic seizures, the patient was unable to form new episodic and declarative memories (i.e., those that can be verbalized) [10]. This finding firmly established a role for the hippocampus in these types of learning and memory, eventually replacing the prevailing inhibition theory of the hippocampus. The inhibition theory had been based on observations of hyperactivity and difficulty learning to inhibit responses following hippocampal damage [11, 12].

An additional theory of hippocampal function was developed in 1971 with O'Keefe's discovery of hippocampal place cells in rats [13]. The intensity of these cells' activity depended on the animal's location within a baited maze. Extensive study was undertaken to identify which environmental cues were used by the animal to recognize its spatial position, and whether activity of the place cells showed spatially selectivity when the animals were placed in a different environment [14, 15]. Navigational problems were observed following hippocampal lesions [16, 17]. Whether the mnemonic and spatial properties are functionally distinct or different facets of the same overarching function was a matter of debate, however, which continues to this day [18–21].

Through most of the twentieth century, theories of hippocampal function relied on evidence from lesion and anatomical studies, plus recordings of electrical activity. The advent of neuroimaging methodologies, particularly functional magnetic resonance imaging (fMRI), allowed hippocampal function to be studied non-invasively in humans. This technique detects regional changes in oxygenated blood resulting from increased neuronal activity, providing the means to identify brain areas based on their functional activity. With the advent of fMRI, investigators could verify in humans the patterns of functional activity observed in non-human animals, adding more complex tasks to further elucidate functional properties.

Early neuroimaging studies examined mean changes in neural activity that differentiated between blocks of time where different tasks were performed, tasks that differed in their cognitive requirements (e.g., memory). Soon, methods were enhanced to identify neural activity during individual trials [22]. Consistent with its theorized mnemonic function, regional increases in hippocampal activity were observed during learning and recall; furthermore, greater activity was observed during those learning trials where a stimulus was presented that was later recalled successfully [23, 24]. Similarly, hippocampal activity during virtual navigation experiments could be correlated with spatial cues [25–27], consistent with its proposed function as a cognitive map. Thus, the two prevalent theories of hippocampal function were both supported. Additional studies described new properties, such as sensitivity to the temporal duration or spatial relationships [28–30], and a role in scene perception and reconstruction [31]. Some interpret these properties as contextual elements required for memory recall [19, 28, 32]; others suggest a more fundamental perceptual role in identifying changes in the environment, which may consequently be incorporated into memories [33, 34]. Differences between these viewpoints are often nuanced. As more information about hippocampal activity has accrued, other roles for the hippocampus have also been suggested, including a role in conscious perception [35–37] and cognitive control [38–40].

The traditional "activation" analysis of fMRI data is patterned on traditional methods for analyzing electrical activity from localized regions of the brain. It assumes all information in a neuron's electrical activity is carried through its frequency of discharge, yet additional information is carried in the hippocampal temporal pattern of activity [41–43]. Cognitive functions also require interactions between neural structures. With the development of connectivity analysis from fMRI data early this century, influences between brain regions could be inferred based on the temporal pattern of neural activity. Early connectivity studies used *functional connectivity*, any of several statistical methods that examines correlations in neural activity between brain regions. Although useful for broadly identifying connections and identifying their abnormalities, the direction of influence in these studies cannot be known with certainty; two regions with correlated activity, for example, might both be influenced from a third region. Methods were soon developed to analyze *effective connectivity*, the influence of one brain region over another.

This chapter will focus on the influence of the hippocampus across a variety of cognitive domains; as such, effective connectivity studies will be emphasized, with particular attention to those that use *psychophysiological interactions* (*PPI*). This form of effective connectivity analysis reveals task-specific influences between regions. Results show a pattern whereby the hippocampus consistently influences activity in cortical areas involved in task performance, including tasks requiring movements, sensory processing, language judgments, and memory. Careful consideration of results and the cognitive requirements of these tasks suggests hippocampal connectivity could play a role in cognitive control, perhaps in parallel with the role of prefrontal cortex in translating thoughts into action.

2. Task-specific connectivity of the hippocampus

2.1 Connectivity with sensorimotor regions during movement tasks

Known hippocampal properties of memory or navigation are not required by common daily movements, such as walking or even "automatic driving" behaviors, so few human studies have examined hippocampal connectivity with motor regions. Initially, those that did examined hippocampal connectivity during sequence learning [44–46], where subjects learn an unfamiliar pattern of finger movements. Hippocampal connectivity was observed with the striatum, suggesting a mnemonic-motor interaction [46], perhaps culminating in striatal-associated movements derived from habits [47, 48].

Hippocampal connectivity with sensorimotor cortex (SMC) was recently studied with PPI during two paced motor tasks, only one of which involved motor learning [49]. For both tasks, subjects were instructed to listen to a 2 Hz metronome for 2 s before initiating movements, then tap the appropriate finger in synchrony with the taps from a metronome. Subjects quickly anticipated the timing of the taps, moving shortly before the sound; thus, cognitive awareness of the expected timing informed motor behavior. During the sequence learning task, the temporal precision and variability of right-handed finger movements improved with repetition; no such learning effects were observed in the repetitive tapping task.

Sensorimotor cortical activation during these tasks (**Figure 1A**) was consistent with previous studies. The sequence learning task was performed with the right hand only, evoking focal activation in left sensorimotor cortex, both in pre- and postcentral gyrus; repetitive tapping was performed with both hands, evoking bilateral sensorimotor activation. Although positive connectivity during sequence

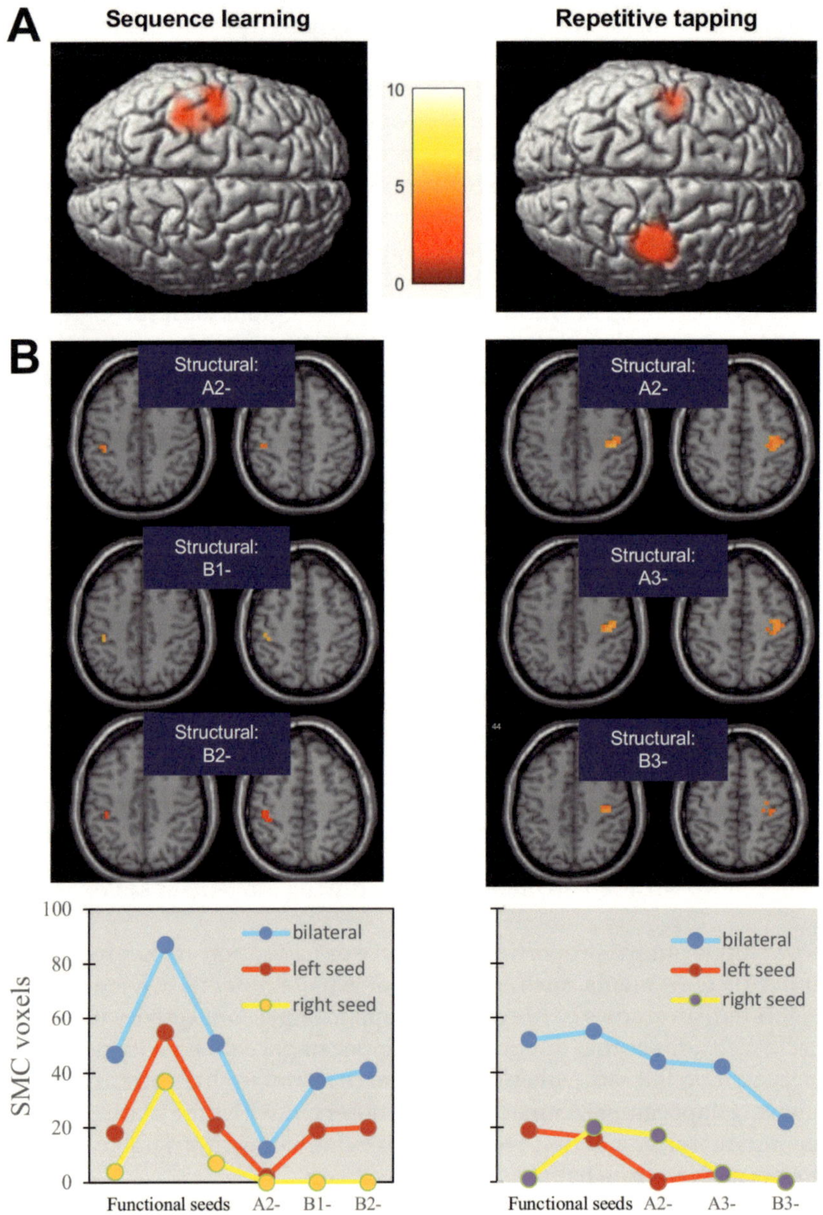

Figure 1.
Sensorimotor activation during performance of motor tasks. (A) Group analysis revealed unilateral activation in left sensorimotor cortex during performance of the unimanual sequence learning task and bilateral activation during the bimanual repetitive tapping task. (B) Inverse connectivity was generated from three structural seeds in both tasks, evident in the left sensorimotor cortex during sequential learning and the right sensorimotor cortex during repetitive tapping. The extent of connectivity was larger using combined ("bilateral") activity from corresponding regions of the left and right hippocampus (bottom). Images are shown in the neurological format (left side of axial images represents left side of brain); structural and functional seeds refer to the method of selecting hippocampal seed regions for connectivity analysis, as described elsewhere [49].

learning was also observed from localized hippocampal regions, inverse (negative) connectivity was more prevalent. During sequence learning, inverse connectivity from central and central-medial regions of the left hippocampus was observed in left SMC (**Figure 1B**, left); the volume of connectivity was slightly improved when examining joint connectivity from the hippocampus in both hemispheres. Inverse connectivity from anterior middle and lateral regions of the hippocampus was observed during repetitive tapping in the right SMC, most evident examining joint connectivity from the hippocampus in both hemispheres (**Figure 1B**, right).

During both tasks, hippocampal connectivity selectively targeted the hand representation within SMC, overlapping the region activated by the task.

2.2 Connectivity with sensory regions during sensory tasks

Sensory information passes into the brain passively through bottom-up processes, but can be enhanced or filtered through top-down processes [50–53]. Top-down processes modify neural responses based on expectations or attentional processes.

Hippocampal influences on cortical processes were examined on tasks that enhanced sensory activation. The Stroop task requires particular attention to colors. On separate trials, words that name colors and cross symbols were presented in colored inks. The word meaning may or may not represent the same color as the ink in which it is written, but the correct behavioral response depends on the ink color. Due to interference from the automatic recognition of the word meaning, subjects require extra time to respond on word trials, especially on mismatch trials; to respond correctly, they must attend to the stimulus color while suppressing the behavioral response suggested by the word meaning. Activation by colored words (vs. colored crosses) was observed in the left inferior frontal gyrus and ventrolateral visual cortex (see **Figure 2A**). Within the activated region, a ventrolateral visual cortex specialized for color [54, 55] showed inverse connectivity from the hippocampus (**Figure 2B**). Despite activation in the left inferior frontal gyrus, an area involved in language function, no hippocampal connectivity was observed there.

Images and sounds that evoke a strong emotional response evoke strong activation in sensory cortices. **Figure 3** shows activation and connectivity associated with a task where subjects were instructed to pay attention to music and images, allowing an emotional response to what they viewed. Pictures were presented from a national database where thousands of subjects had rated images for the intensity and sign of their affective response [56]; harsh dissonant music accompanied negative images, upbeat classical music accompanied positive images, and bland jazz music accompanied neutral images, interspersed between positive and negative images. **Figure 3** shows brain activity evoked by negative stimuli. **Figure 3A** shows bilateral activation in visual cortex along the calcarine sulcus, plus auditory association cortex within the superior temporal gyrus; a similar pattern of activation was observed for positive stimuli (not shown). Hippocampal connectivity was not observed in visual cortex, whereas inverse connectivity was observed bilaterally in the activated region of the superior temporal gyrus (**Figure 3B**).

Tactile brain activation was tested by rubbing the arms of thirty-five patients with brain tumors evaluated during pre-operative planning. Patients were instructed to attend to the spatial pattern of tactile stimuli, which were applied bilaterally; analysis was carried out separately for those patients with tactile impairments on the left vs. right sides. Bilateral activation was generated in the postcentral gyrus, weaker in the sensory cortex contralateral to the sensory deficit (not shown). Hippocampal connectivity was absent.

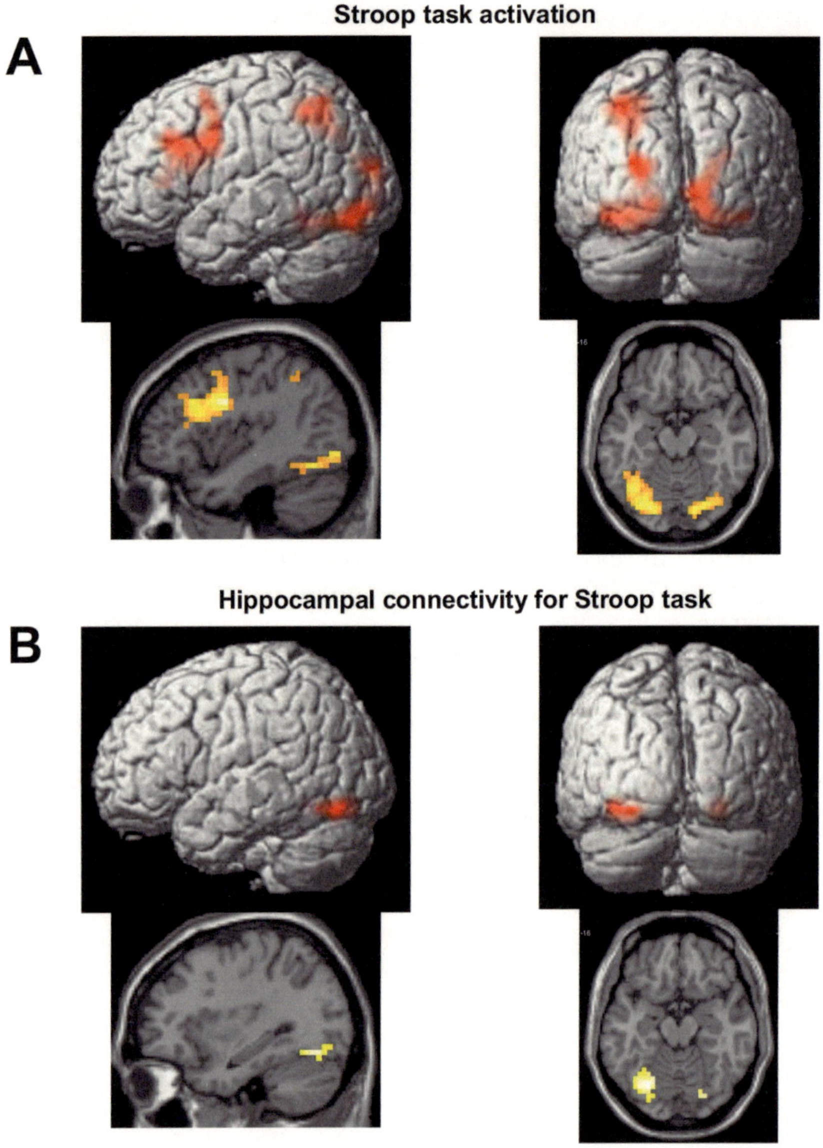

Figure 2.
Activation and hippocampal connectivity during performance of Stroop task. (A) Activation was observed in the left inferior frontal gyrus (a language area), the left inferior parietal lobe, and bilateral occipital cortex, extending into the fusiform gyrus (visual areas). (B) Inverse connectivity from bilateral seeds in the center of the hippocampus was observed in fusiform regions associated with color processing.

As illustrated above, hippocampal connectivity was never observed in a primary sensory region, but was observed in activated regions of sensory association cortex (for example, the activated color association cortex during the Stroop task and auditory association cortex during dissonant music). This pattern of results has functional implications. Patterns of visual cortex activity are constrained by attentional processes and cognitive expectations [57], and the hippocampal mechanism of pattern completion reflects cognitive expectations [58]. Visual responses in the hippocampus are retinotopic, suggesting their joint function in sensations and memory [59]. Visual and auditory areas specialized for language also receive hippocampal input (as shown in the next section). This pattern of results suggests consistent hippocampal cognitive input to sensory areas that extract features relevant to task performance.

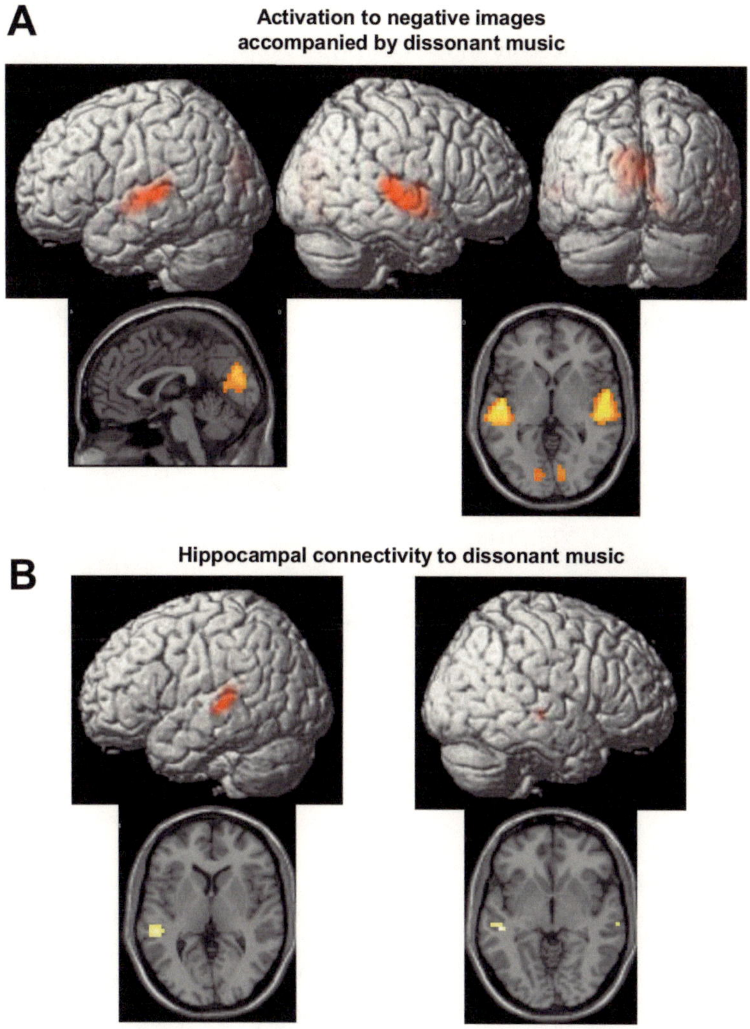

Figure 3.
Activation and hippocampal connectivity during presentation of multisensory emotionally-charged stimuli. (A) Activation was observed bilaterally in the occipital cortex along the calcarine gyrus and the superior temporal gyrus. Emotionally-charged photographic images accompanied by dissonant music were contrasted with neutral images of furniture, faces, and scenery. (B) Hippocampal connectivity was limited to auditory association cortex within the superior temporal gyrus.

2.3 Connectivity with language regions during language tasks

The language network consists of interconnected brain regions that vary in linguistic properties. The left inferior frontal gyrus (Broca's area) is typically active during all language tasks, although subregions have been identified with various linguistic functions [60, 61]. Occipital and temporal regions are specialized for processing specific linguistic components [62–67]. The default mode network typically shows decreased activity during language judgment tasks, yet the magnitude of its activation and connectivity with language areas can be correlated with performance accuracy [68, 69].

Representative activity during three language tasks is shown in **Figure 4**. **Figure 4A** shows results from both a t-test analysis (representing positive activation, yellow) and an anova F-test (red); overlap appears in orange. The t-test analysis shows the traditional language network, including the left inferior frontal gyrus, middle and adjacent superior temporal gyrus, and the fusiform gyrus. The F-test analysis additionally shows deactivation in the default mode network, including the precuneus, angular gyrus, and ventromedial prefrontal cortex. Regions with hippocampal connectivity during the auditory version of these tasks is shown in **Figure 4B**, both for adults (yellow) and children (red). In the orthography task, *inverse* connectivity was observed for children in the left fusiform gyrus and the posterior default mode network; all other regions reflect *positive* connectivity. A larger area of connectivity was observed in adults for phonology

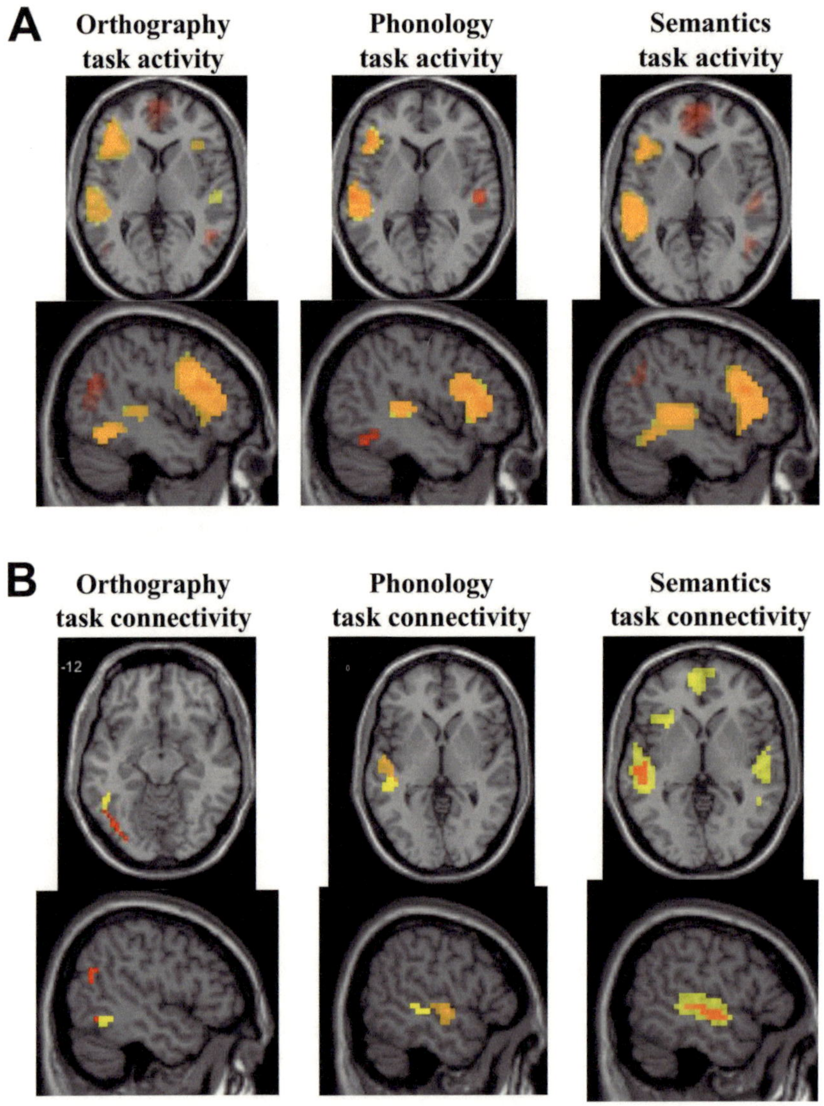

Figure 4.
Activation and hippocampal connectivity during language tasks. (A) Activation during language tasks was evaluated both with an F-test (red) and t-test (yellow). Areas identified from the t-test are traditionally associated with language activation, including the left inferior frontal gyrus, middle/superior temporal gyrus, and fusiform gyrus; the F-test additionally demonstrated areas in the default mode network (precuneus, angular gyrus, and ventromedial prefrontal cortex). (B) Hippocampal connectivity varied across different language tasks. During the orthography task, hippocampal connectivity in the fusiform gyrus was more anterior in children than adults, who also showed connectivity in the angular gyrus. Hippocampal connectivity in the phonology and semantic tasks overlapped in the temporal gyrus, extending further posterior during semantics; the semantics task additionally showed connectivity in the left inferior frontal gyrus and ventromedial prefrontal cortex.

and semantic tasks, encompassing the area of connectivity observed in children. In addition, adults showed connectivity in the left insula/inferior frontal gyrus (Broca's area) in the semantics task, as well as ventromedial prefrontal cortex of the default mode network.

These three language tasks varied only in the linguistic judgment required for accurate performance [70, 71]. Three words were presented sequentially; the required response depended on the rule designated for that task (the third word must be spelled the same, rhyme, or be related in meaning to either of the two previous words). Hippocampal activity likely reflected its memory for the first two words, consistent with its mnemonic function, yet its connectivity with language areas depended on the task requirements. The three language tasks preferentially activated different areas in the language network (**Figure 5A**): fusiform gyrus

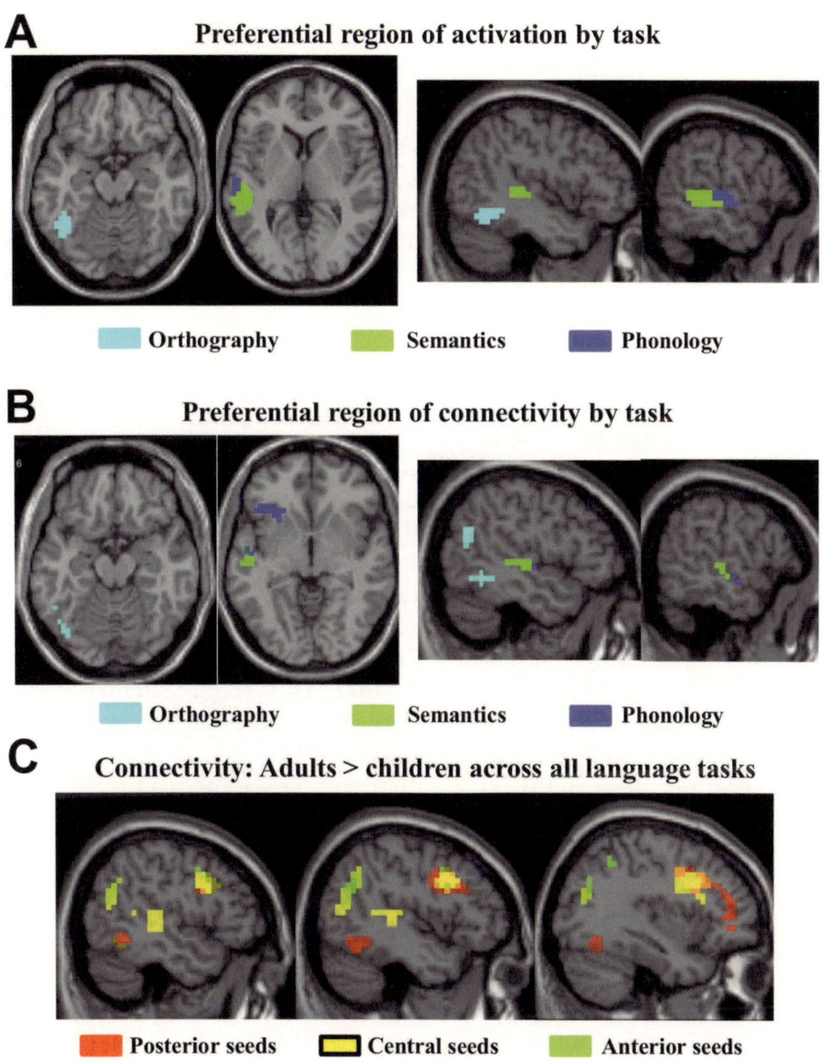

Figure 5.
Language task selectivity and developmental changes in hippocampal connectivity. (A) Task-preferential activation was observed in the left fusiform gyrus (orthography), posterior middle/superior temporal gyrus (semantics), and adjacent superior temporal gyrus (phonology). (B) Hippocampal connectivity showed the same task-dependent pattern. (C) Developmental increases in hippocampal connectivity were observed in most language areas (left fusiform, posterior middle/superior temporal gyrus, inferior frontal gyrus), plus part of the default mode network (precuneus, angular gyrus). Different regions of the hippocampus showed developmental increases with different cortical areas.

for orthography, superior temporal gyrus for phonology, and posterior middle temporal gyrus for semantics. These same areas showed task-specific connectivity from the hippocampus (**Figure 5B**). Language deficits are associated with abnormal activity or connections in these areas [72, 73]; thus, hippocampal connectivity altered the activity in those language areas necessary for performing the task.

Hippocampal connectivity with language areas increased through adulthood, with different hippocampal regions showing developmental increases in connectivity with different cortical areas (**Figure 5C**). Areas with increased connectivity included the fusiform and posterior middle temporal regions (associated with spelling and semantics, respectively), but also the inferior frontal gyrus and parts of the default mode network. Developmental changes have also been observed within the language network, both in activation [71, 74–76] and connectivity [77–80]. Developmental changes have been tied to changes in language skills [81, 82]. Developmental changes in hippocampal connectivity may reflect cognitive changes associated with these language skills.

2.4 Connectivity with prefrontal regions during memory tasks

Different types of memory are often differentiated by interactions between different brain regions. Many memories are believed to involve the hippocampus, either during memory formation, the search for a specific memory ("construction" for autobiographical memories), its elaboration during recall, or updating {"integrating") memories to incorporate new content.

In autobiographical memories, posterior hippocampal regions interact with visual perceptual areas [83, 84]; during the construction of these memories and imagined future events, anterior hippocampal regions show increased connectivity with prefrontal regions [83]. New information inconsistent with a previous schema changes hippocampal-prefrontal connectivity [85]. Successful integration of recalled information in an inference task results in enhanced hippocampal theta power, plus coherence in medial prefrontal cortex [86], suggesting a directional flow of information from the hippocampus to prefrontal areas. During a working memory task, hippocampal activity precedes frontal activity during successful trials [86], reflecting successful retrieval and suggesting a directional flow of information from the hippocampus to prefrontal cortex. In combined EEG-fMRI recordings, recollection-specific theta-alpha (4–13 Hz) effects are correlated with increases in hippocampal connectivity with the PFC and the striatum, areas that have been linked repeatedly to retrieval success [87, 88].

Information also flows in the reverse direction. Ventromedial prefrontal cortex drives the hippocampus during the generation and processing of mismatch signals; the hippocampus then integrates this information into a new schema, modifying existing memories [89]. Elaboration of emotional autobiographical memories generates connectivity from ventromedial prefrontal cortex to the hippocampus, with greater connectivity generated during highly emotionally arousing events than those with neutral or positive affect [83]. Prefrontal feedback may thus reinforce the strength of hippocampal activity based on emotional content, explaining why emotionally-charged events are more likely to be remembered [90, 91]. Bidirectional interactions between the hippocampus and medial prefrontal cortex also play a role in working, episodic, and spatial memory [92, 93], with dysfunction in these pathways likely contributing to psychiatric disorders [94, 95]. The pattern of information flow suggests that the context of ongoing experience (the schema) is required to retrieve relevant memories, allowing patterns of neural activity from the original event to be recreated in sensory association cortices [96, 97].

In summary, the flow of information between the hippocampus and prefrontal cortex is bidirectional during memory-related tasks: the hippocampus provides contextual information when novel stimuli or patterns appear, with feedback from the prefrontal cortex resetting the contextual schema in perceptual areas that provide input to the hippocampus.

3. Implications for cognitive control

3.1 Hippocampal properties are consistent with a role in cognitive control during volitional movements

Cognitive control has variously been defined as a psychological construct for the coordination of thoughts and actions under conditions of uncertainty [98], the collective processes that organize different thoughts and memories, allowing the separation of currently relevant and irrelevant information [40], brain processes involved in regulating behavior according to internal goals or plans [38], and the ability to coordinate multiple streams of information to prevent confusion and select appropriate behavioral responses, especially when presented with competing alternatives [39]. Cognitive control processes allow us to efficiently process information and generate appropriate responses.

The essence of cognitive control is that neural processes involved in cognitive (psychological) processes act upon those regions of the brain needed to translate our thoughts into action. In this sense, "action" includes processes involved in making decisions, accessing memories, attentional control, response inhibition, and mental computations – i.e., any process that can potentially result in changes in behavior due to ongoing mental activity. Early explorations used tasks that unambiguously require cognition, such as:

- the Wisconsin card sorting task. A rule has to be identified to select the correct card for a reward, then the rule changes.
- the Stroop task. The subject's response must be based on ink color, ignoring the meaning of presented words that refer to a different color.
- the n-back task. Letters are presented sequentially, and the subject indicates whether the currently-presented letter matches the letter appearing n-stimuli previously.

Neural activity that correlates with these task behaviors is observed in prefrontal cortex [99–102], and prefrontal lesions in animals [103, 104] or humans [105] impair task performance. Neurological disorders associated with impaired frontal function, such as schizophrenia, also show impairments on these tasks [106]. The prefrontal cortex exerts top-down influences on sensory areas by functioning as a filtering mechanism that biases bottom-up sensory information toward the optimal response for a given context [107, 108]. This feedback loop may also be involved in memory recall, since the act of remembering evokes activity in the same sensory areas as the original event [96, 97]. The accumulated evidence supports a role of the prefrontal cortex in cognitive control.

Nonetheless, not all aspects of complex cognition benefit from prefrontal regulation, and the prefrontal cortex is not the sole source of cognitive control [109–111]. Indeed, the role for the prefrontal cortex in some functions may be limited. Prefrontal influence on activity in the primary motor cortex is indirect via dorsal premotor cortex when learning to perform sequential finger movements, and absent during repetitive movements [112–114]. However, repetitive finger movements paced by a metronome do reflect cognitive control, as movements soon anticipate the auditory cue [49]. Under these conditions, the hippocampus provides connectivity selectively to the sensorimotor cortex hand representation, as shown in **Figure 1**.

An unequivocal set of criteria for a role in cognitive control has never been established, most studies relying on correlations between cognitive task performance and neural activity. In proposing minimal criteria for the cognitive control of movements [38], Burman noted an analogy between the skeletal movement system and the frontal eye field (FEF), which plays a critical role in volitional eye movements [115, 116]. Modifiable by cognitive influences, FEF cells only have three response properties, providing the basis for his proposed criteria:

1. Neural activity must be tied to a cognitive/volitional state of consciousness.

2. Cognitive influences must be selective for the time period required for task performance (*temporal selectivity*).

3. Cognitive influences must be selective for the spatial region to be acted upon (*spatial selectivity*).

Hippocampal connectivity with sensorimotor cortex during the repetitive tapping task arguably meets these criteria. Topographical connectivity maps were identified from single-voxel functional seeds that differentiated between movements of adjacent fingers [38]. Using the finger representations identified from an earlier study [117], the intensity of connectivity from each functional seed was then compared statistically across time periods for movement of each finger. The criteria for cognitive control were met:

1. The repetitive tapping task involved a *volitional state of consciousness*. The timing of movements resulted from a cognitive schema, as movements anticipated the auditory cue in this task, whereas the hippocampus is unresponsive during implicit learning of movements [118].

2. At each finger representation, maximal connectivity was observed during the time period when the represented finger was moving (*temporal selectivity*). Connectivity tapered during the time period when the adjacent finger moved, with residual activity likely resulting from coupled movements between adjacent fingers.

3. Throughout the task, maximal connectivity was observed at the representation of the finger that was currently moving (*spatial selectivity*). With the fingers and response pad at a fixed location, movement of each finger represented movement in a specific region in space.

The extent to which these results can be generalized is limited, as the spatial area covered by finger movements was restricted, only one temporal pattern of movements was tested, and only finger movements were involved. Nonetheless, known hippocampal properties are consistent with these conclusions: hippocampal function is associated with conscious states such as declarative memories, and the hippocampus is sensitive to the timing of events as well as their spatial properties. Exploring a wider range of spatial regions involved in physical manipulation, as well as varied durations of movements, could more fully delineate the extent of its cognitive control over volitional movements.

3.2 Evidence for cognitive control suggested by prefrontal and hippocampal connectivity studies

Figure 6 summarizes cortical connections between areas involved in sensory, language, memory, and motor functions. Sensory input from primary sensory cortices passes to association cortex for feature extraction (yellow arrows), then to higher centers involved in language, memory formation and cognition. The higher centers have bidirectional connections (blue arrows); in addition, the hippocampus and prefrontal cortex modify activity in sensory, language, and motor areas (red arrows). As described below, the red arrows represent candidates for cognitive control.

Cognitive control requires influence from a higher center to modify neural activity in those areas required to perform a task. The prefrontal cortex and hippocampus [40] have both been suggested to play such a role, and both have connections appropriate for a role in cognitive control over sensory input, language, and memory (as shown in **Figure 6**). Connectivity from prefrontal cortex to motor areas is indirect and limited to sequential movement learning tasks; otherwise, there is little in the pattern of connectivity to differentiate between these candidates. The hippocampus and prefrontal cortex show strong interactions, suggesting they may often work jointly to exert cognitive control. Such redundancy would have an evolutionary advantage, since damage to either area by itself will be less crippling. As shown in **Figure 6**, hippocampal effects on sensory areas may also be indirectly mediated through prefrontal cortex. This could explain a number of curious findings, such as why the hippocampus is needed for memory formation (to provide contextual information to prefrontal cortex) but not for memory recall (when prefrontal cortex provides memory recall by reactivating sensory areas involved in sensory

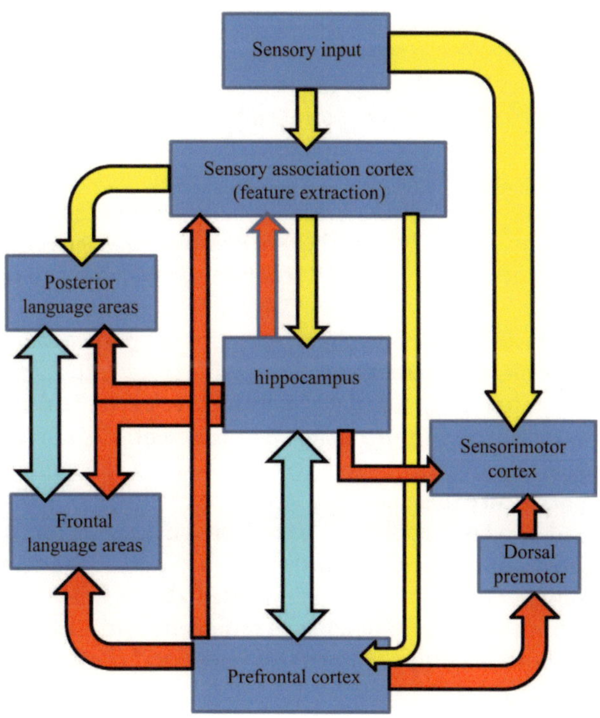

Figure 6.
Summary diagram of cortical connections. Sensory pathways are colored yellow, modulatory feedback pathways red, and bidirectional connections blue. Red pathways in this summary originate from prefrontal cortex and the hippocampus, which are both likely to play a role in cognitive control, informed by their mutual connections and feedback.

perception). Such redundancy can also account for residual cognitive abilities during neurological disorders that disrupt function in either region.

A recent study compared functional connectivity during tasks and the resting state condition, the latter representing the intrinsic architecture of the brain [119]. Small but consistent changes were observed across dozens of task states, suggesting both task-specific and task-general network changes. Appearing within the resting state network, the hippocampus (but not prefrontal cortex) accounted for most variance in connectivity across all tasks. This finding suggests that the hippocampus, unique from prefrontal cortex, plays a primary role in regulating task behavior.

3.3 Testing for cognitive control in the hippocampus and prefrontal cortex

Regardless of the brain region studied, the same criteria can be applied to establish a role in cognitive control. It was previously noted that early studies used tasks that unambiguously required cognition to perform, such as the Stroop and n-back tasks; prefrontal cortex was implicated because response properties correlated with task performance. This approach addresses the first of the three criteria proposed above: the neural activity is tied to a cognitive or volitional state of consciousness. For the hippocampus, the need to relate hippocampal activity to a cognitive or volitional state may not initially be apparent due to its association with declarative memory, yet hippocampal activity has also been reported during the formation of implicit memories [120–122]. Until we know the functional role of hippocampal activity during implicit learning conditions (decision-making? context? association pairings?), we cannot assume that hippocampal activity necessarily reflects a volitional state of consciousness. Behaviors acquired through hippocampus from repeated stimulus–response associations, for example, require little thought and may ultimately be mediated by the cerebellum and striatum [48, 123–125].

Involvement in cognitive processes does not in itself indicate a role in cognitive control. The left inferior frontal gyrus (Broca's area) was activated by words in the Stroop task, for example, yet the behavioral response suggested by the meaning of conflicting words had to be suppressed for accurate performance. To play a role in cognitive control, brain regions involved in cognition must act upon brain areas required to achieve the goal of a task. Effective connectivity tools provide a method to study such effects, particularly useful when demonstrable effects are task-specific. (This is the advantage of the PPI technique, which is both directional and task-specific).

Such a task-specific influence must be relevant to task performance. This is the purpose of the temporal- and spatial-selectivity criteria suggested above: task performance is always delimited in time, and typically involve perceptual stimuli or volitional movements appearing within a spatial environment. Cognitive processes such as emotional associations may not be invariably linked to a concrete stimulus, yet the provocative stimulus in an experiment can still be spatially delimited. Any additional constraints imposed by a task should also be reflected in a signal for cognitive control.

A role in cognitive control can be confirmed when loss of the control signal results in the inability to perform the task. As noted above, however, the behavioral deficit will not be complete unless signals are removed from all areas involved, which may include both the hippocampus and prefrontal cortex.

4. Conclusions

The hippocampus provides task-specific influences on sensory, motor, language, and mnemonic areas of the brain. Detailed analysis in sensorimotor regions during motor tasks showed a pattern of connectivity consistent with the requirements for cognitive control; connectivity patterns across tasks suggest a joint role for the hippocampus and prefrontal cortex. Criteria and suggestions for further evaluation of cognitive control from both regions are considered.

Acknowledgements

The author wishes to thank Samuel Cohan for examining hippocampal connectivity with sensory systems and Donald J. Bolger for helpful discussions on the literature.

Author details

Douglas D. Burman
NorthShore University HealthSystem, Evanston, IL, USA

*Address all correspondence to: dburman2@northshore.org

References

[1] Finger S. Origins of neuroscience: A history of explorations into brain function: Oxford University Press US; 2001.

[2] Papez JW. A proposed mechanism of emotion. Archives of Neurology and Psychiatry 1937;38:725-743.

[3] Swanson LW. The hippocampus and the concept of the limbic system. In: Seifert W, editor. Neurobiology of the Hippocampus. London, UK: Academic; 1983.

[4] Klüver H, Bucy PC. An analysis of certain effects of bilateral temporal lobectomy in the rhesus monkey, with special reference to psychic blindness. Journal of Psychology 1938;5(1):33-54.

[5] Lilly R, Cummings JL, Benson DF, Frankel M. The human Klüver-Bucy syndrome. Neurology 1983;33(9):1141-1141.

[6] Marlowe WB, Mancall EL, Thomas JJ. Complete Klüver-Bucy syndrome in man. Cortex 1975;11(1):53-59.

[7] Green JD, Arduini AA. Hippocampal electrical activity in arousal. Journal of Neurophysiology 1954;17(6):533-557.

[8] Lashley KS. In Search of the Engram: Academic Press; 1950.

[9] Eichenbaum H. Memory, Amnesia, and the Hippocampal System: MIT Press; 1993.

[10] Scoville WB, Milner B. Loss of recent memory after bilateral hippocampal lesions. J Neurol Neurosurg Psychiatry 1957;20(1):11-21.

[11] Douglas RJ. The hippocampus and behavior. Psychological Bulletin 1967;67(6):416.

[12] Kimble DP. Hippocampus and internal inhibition. Psychological Bulletin 1968;70(5):285.

[13] O'Keefe J, Dostrovsky J. The hippocampus as a spatial map: preliminary evidence from unit activity in the freely-moving rat. Brain Research 1971.

[14] Muller R. A quarter of a century of place cells. Neuron 1996;17(5):813-822.

[15] Poucet B, Save E, Lenck-Santini P-P. Sensory and memory properties of hippocampal place cells. Reviews in the Neurosciences 2000;11(2-3):95-112.

[16] Hollup SA, Kjelstrup KG, Hoff J, Moser M-B, Moser EI. Impaired recognition of the goal location during spatial navigation in rats with hippocampal lesions. Journal of Neuroscience 2001;21(12):4505-4513.

[17] Whishaw IQ, McKenna JE, Maaswinkel H. Hippocampal lesions and path integration. Current Opinion in Neurobiology 1997;2(7):228-234.

[18] Eichenbaum H, Cohen NJ. Can we reconcile the declarative memory and spatial navigation views on hippocampal function? Neuron 2014;83(4):764-770.

[19] Eichenbaum H. On the integration of space, time, and memory. Neuron 2017;95(5):1007.

[20] Rolls ET. Functions of the primate hippocampus in spatial processing and memory. In: Neurobiology of Comparative Cognition: Psychology Press; 2014. p. 359-382.

[21] Smith DM, Mizumori SJY. Hippocampal place cells, context, and episodic memory. Hippocampus 2006;16(9):716-729.

[22] Hopfinger JB, Büchel C, Holmes AP, Friston KJ. A study of analysis parameters that influence the sensitivity of event-related fMRI analyses. NeuroImage 2000;11(4):326-333.

[23] Hannula DE, Ranganath C. Medial temporal lobe activity predicts successful relational memory binding. Journal of Neuroscience 2008;28(1):116-24.

[24] Jenkins LJ, Ranganath C. Prefrontal and medial temporal lobe activity at encoding predicts temporal context memory. Journal of Neuroscience 2010;30(46):15558-15565.

[25] Hirshhorn M, Grady C, Rosenbaum RS, Winocur G, Moscovitch M. The hippocampus is involved in mental navigation for a recently learned, but not a highly familiar environment: a longitudinal fMRI study. Hippocampus 2012;22(4):842-852.

[26] Ohnishi T, Matsuda H, Hirakata M, Ugawa Y. Navigation ability dependent neural activation in the human brain: an fMRI study. Neuroscience Research 2006;55(4):361-369.

[27] Rodriguez PF. Neural decoding of goal locations in spatial navigation in humans with fMRI. Human brain mapping 2010;31(3):391-397.

[28] Barnett AJ, O'neil EB, Watson HC, Lee ACH. The human hippocampus is sensitive to the durations of events and intervals within a sequence. Neuropsychologia 2014;64:1-12.

[29] Libby LA, Hannula DE, Ranganath C. Medial temporal lobe coding of item and spatial information during relational binding in working memory. Journal of Neuroscience 2014;34:14233-14242.

[30] Olsen RK, Moses SN, Riggs L, Ryan JD. The hippocampus supports multiple cognitive processes through relational binding and comparison. Frontiers in Human Neuroscience 2012;6:146.

[31] Zeidman P, Mullally SaL, Maguire EA. Constructing, perceiving, and maintaining scenes: hippocampal activity and connectivity. Cerebral Cortex 2015;25(10):3836-3855.

[32] Eichenbaum H. Time cells in the hippocampus: a new dimension for mapping memories. Nature Reviews Neuroscience 2014;15(11):732-744.

[33] Howard MW, Eichenbaum H. Time and space in the hippocampus. Brain Research 2015;1621:345-354.

[34] MacDonald CJ. Prospective and retrospective duration memory in the hippocampus: is time in the foreground or background? Phil. Trans. R. Soc. B 2014;369(1637):20120463.

[35] Behrendt R-P. Contribution of hippocampal region CA3 to consciousness and schizophrenic hallucinations. Neuroscience & Biobehavioral Reviews 2010;34(8):1121-1136.

[36] Behrendt R-P. Hippocampus and consciousness. Reviews in the Neurosciences 2013;24(3):239-266.

[37] Behrendt R-P. Conscious experience and episodic memory: hippocampus at the crossroads. Frontiers in Psychology 2013;4:304.

[38] Burman DD. Hippocampal connectivity with sensorimotor cortex during volitional finger movements. II. Spatial and temporal selectivity. bioRxiv 2018;doi: https://doi.org/10.1101/479436.

[39] Kelemen E, Fenton AA. Dynamic grouping of hippocampal neural activity during cognitive control of two spatial frames. PLoS Biology 2010;8(6):e1000403.

[40] Kelemen E, Fenton AA. Coordinating different representations in the hippocampus. Neurobiology of Learning and Memory 2016;129:50-59.

[41] Hsieh L-T, Gruber MJ, Jenkins LJ, Ranganath C. Hippocampal activity patterns carry information about objects in temporal context. Neuron 2014;81(5):1165-1178.

[42] Klemm WR, Sherry CJ. Do neurons process information by relative intervals in spike trains? Neuroscience and Biobehavioral Reviews 1982;6(4):429-437.

[43] Somogyi P, Klausberger T. Defined types of cortical interneurone structure space and spike timing in the hippocampus. Journal of Physiology 2005;562(1):9-26.

[44] Albouy G, Sterpenich V, Vandewalle G, Darsaud A, Gais S, Rauchs G, et al. Interaction between hippocampal and striatal systems predicts subsequent consolidation of motor sequence memory. PLoS One 2013;8(3):e59490.

[45] Albouy G, Sterpenich V, Balteau E, Vandewalle G, Desseilles M, Dang-Vu T, et al. Both the hippocampus and striatum are involved in consolidation of motor sequence memory. Neuron 2008;58(2):261-272.

[46] Fernandez-Seara MA, Aznarez-Sanado M, Mengual E, Loayza FR, Pastor MA. Continuous performance of a novel motor sequence leads to highly correlated striatal and hippocampal perfusion increases. Neuroimage 2009;47(4):1797-1808.

[47] Albouy G, Fogel S, King BR, Laventure S, Benali H, Karni A, et al. Maintaining vs. enhancing motor sequence memories: Respective roles of striatal and hippocampal systems. Neuroimage 2015;108:423-434.

[48] Packard MG, Knowlton BJ. Learning and memory functions of the basal ganglia. Annual Review of Neuroscience 2002;25:563-93.

[49] Burman DD. Hippocampal connectivity with sensorimotor cortex during volitional finger movements: Laterality and relationship to motor learning. PloS One 2019;14(9):e0222064.

[50] Gilbert CD, Li W. Top-down influences on visual processing. Nature Reviews Neuroscience 2013;14(5):350-363.

[51] Ranganath C, D'Esposito M. Directing the mind's eye: prefrontal, inferior and medial temporal mechanisms for visual working memory. Current Opinion in Neurobiology 2005;15(2):175-82.

[52] Kok P, Turk-Browne NB. Associative prediction of visual shape in the hippocampus. Journal of Neuroscience 2018;38(31):6888-6899.

[53] Langner R, Kellermann T, Boers F, Sturm W, Willmes K, Eickhoff SB. Modality-specific perceptual expectations selectively modulate baseline activity in auditory, somatosensory, and visual cortices. Cerebral Cortex 2011;21(12):2850-2862.

[54] Conway BR. Color vision, cones, and color-coding in the cortex. The Neuroscientist 2009;15(3):274-290.

[55] Fritch HA, Thakral PP, Slotnick SD, Ross RS. Distinct patterns of hippocampal activity associated with color and spatial source memory. Hippocampus 2021.

[56] Lang P, Bradley MM. The International Affective Picture System (IAPS) in the study of emotion and attention. Handbook of Emotion Elicitation and Assessment 2007;29.

[57] Cördova NI, Tompary A, Turk-Browne NB. Attentional modulation of background connectivity between ventral visual cortex and the

medial temporal lobe. Neurobiology of Learning and Memory 2016;134:115-122.

[58] Hindy NC, Ng FY, Turk-Browne NB. Linking pattern completion in the hippocampus to predictive coding in visual cortex. Nature Neuroscience 2016;19(5):665-667.

[59] Knapen T. Topographic connectivity reveals task-dependent retinotopic processing throughout the human brain. Proceedings of the National Academy of Sciences 2021;118(2):e2017032118.

[60] Katzev M, Tüscher O, Hennig J, Weiller C, Kaller CP. Revisiting the functional specialization of left inferior frontal gyrus in phonological and semantic fluency: the crucial role of task demands and individual ability. Journal of Neuroscience 2013;33(18):7837-7845.

[61] Klaus J, Hartwigsen G. Dissociating semantic and phonological contributions of the left inferior frontal gyrus to language production. Human Brain Mapping 2019;40(11):3279-3287.

[62] Booth JR, Burman DD, Meyer JR, Gitelman DR, Parrish TB, Mesulam MM. Modality independence of word comprehension. Human Brain Mapping 2002;16(4):251-61.

[63] Booth JR, Burman DD, Meyer JR, Gitelman DR, Parrish TB, Mesulam MM. Functional anatomy of intra- and cross-modal lexical tasks. Neuroimage 2002;16(1):7-22.

[64] Vigneau M, Beaucousin V, Herve PY, Duffau H, Crivello F, Houde O, et al. Meta-analyzing left hemisphere language areas: phonology, semantics, and sentence processing. Neuroimage 2006;30(4):1414-32.

[65] Xue G, Chen C, Jin Z, Dong Q. Language experience shapes fusiform activation when processing a logographic artificial language: an fMRI training study. Neuroimage 2006;31(3):1315-26.

[66] Cohen L, Lehericy S, Chochon F, Lemer C, Rivaud S, Dehaene S. Language-specific tuning of visual cortex? Functional properties of the Visual Word Form Area. Brain 2002;125(Pt 5):1054-69.

[67] Vartiainen J, Parviainen T, Salmelin R. Spatiotemporal convergence of semantic processing in reading and speech perception. Journal of Neuroscience 2009;29(29):9271-80.

[68] Evans M, Krieger-Redwood K, Alam TRJG, Smallwood J, Jefferies E. Controlled semantic summation correlates with intrinsic connectivity between default mode and control networks. Cortex 2020;129:356-375.

[69] Gordon EM, Laumann TO, Marek S, Raut RV, Gratton C, Newbold DJ, et al. Default-mode network streams for coupling to language and control systems. Proceedings of the National Academy of Sciences 2020;117(29):17308-17319.

[70] Booth JR, Burman DD, Meyer JR, Gitelman DR, Parrish TB, Mesulam MM. Relation between brain activation and lexical performance. Human Brain Mapping 2003;19(3):155-69.

[71] Booth JR, Burman DD, Van Santen F, Harasaki Y, Gitelman DR, Parrish TB, et al. Developmental differences in brain systems for reading. Hrvatska Revija Za Rehabilitacijska Istrazivanja (Croatia Early Communication and Language Development) 2001;37(1) 37-52.

[72] Vannest J, Karunanayaka PR, Schmithorst VJ, Szaflarski JP, Holland SK. Language networks in children: evidence from functional MRI studies. American Journal of Roentgenology 2009;192(5):1190-1196.

[73] Vansteensel MJ, Selten IS, Charbonnier L, Berezutskaya J, Raemaekers MAH, Ramsey NF, et al. Reduced brain activation during spoken language processing in children with developmental language disorder and children with 22q11. 2 deletion syndrome. Neuropsychologia 2021:107907.

[74] Booth JR, Burman DD, Meyer JR, Gitelman DR, Parrish TB, Mesulam MM. Development of brain mechanisms for processing orthographic and phonologic representations. Journal of Cognitive Neuroscience 2004;16(7):1234-49.

[75] Gaillard WD, Sachs BC, Whitnah JR, Ahmad Z, Balsamo LM, Petrella JR, et al. Developmental aspects of language processing: fMRI of verbal fluency in children and adults. Human Brain Mapping 2003;18(3):176-85.

[76] Cone NE, Burman DD, Bitan T, Bolger DJ, Booth JR. Developmental changes in brain regions involved in phonological and orthographic processing during spoken language processing. Neuroimage 2008;41(2):623-35.

[77] Bitan T, Burman DD, Lu D, Cone NE, Gitelman DR, Mesulam MM, et al. Weaker top-down modulation from the left inferior frontal gyrus in children. Neuroimage 2006;33(3):991-8.

[78] Bitan T, Cheon J, Lu D, Burman DD, Gitelman DR, Mesulam MM, et al. Developmental changes in activation and effective connectivity in phonological processing. Neuroimage 2007;38(3):564-575.

[79] Bitan T, Cheon J, Lu D, Burman DD, Booth JR. Developmental increase in top-down and bottom-up processing in a phonological task: an effective connectivity, fMRI study. Journal of Cognitive Neuroscience 2009;21(6):1135-1145.

[80] Booth JR, Mehdiratta N, Burman DD, Bitan T. Developmental increases in effective connectivity to brain regions involved in phonological processing during tasks with orthographic demands. Brain Research 2008;1189:78-89.

[81] Burman DD, Minas T, Bolger DJ, Booth JR. Age, sex, and verbal abilities affect location of linguistic connectivity in ventral visual pathway. Brain and Language 2013;124(2):184-193.

[82] Lidzba K, Schwilling E, Grodd W, Krägeloh-Mann I, Wilke M. Language comprehension vs. language production: age effects on fMRI activation. Brain and language 2011;119(1):6-15.

[83] Campbell KL, Madore KP, Benoit RG, Thakral PP, Schacter DL. Increased hippocampus to ventromedial prefrontal connectivity during the construction of episodic future events. Hippocampus 2018;28(2):76-80.

[84] McCormick C, St-Laurent M, Ty A, Valiante TA, McAndrews MP. Functional and effective hippocampal-neocortical connectivity during construction and elaboration of autobiographical memory retrieval. Cerebral Cortex 2015;25(5):1297-1305.

[85] Bein O, Reggev N, Maril A. Prior knowledge influences on hippocampus and medial prefrontal cortex interactions in subsequent memory. Neuropsychologia 2014;64:320-330.

[86] Liu T, Bai W, Xia M, Tian X. Directional hippocampal-prefrontal interactions during working memory. Behavioural Brain Research 2018;338:1-8.

[87] Herweg NA, Apitz T, Leicht G, Mulert C, Fuentemilla Garriga L, Bunzeck N. Theta-alpha oscillations bind the hippocampus, prefrontal cortex, and striatum during recollection: Evidence from

simultaneous EEG-fMRI. Journal of Neuroscience 2016;36(12):3579-3587.

[88] Wolf RC, Vasic N, Sambataro F, Höse A, Frasch K, Schmid M, et al. Temporally anticorrelated brain networks during working memory performance reveal aberrant prefrontal and hippocampal connectivity in patients with schizophrenia. Progress in Neuro-Psychopharmacology and Biological Psychiatry 2009;33(8):1464-1473.

[89] Garrido MI, Barnes GR, Kumaran D, Maguire EA, Dolan RJ. Ventromedial prefrontal cortex drives hippocampal theta oscillations induced by mismatch computations. Neuroimage 2015;120:362-370.

[90] Talmi D. Enhanced emotional memory: Cognitive and neural mechanisms. Current Directions in Psychological Science 2013;22(6):430-436.

[91] Humphreys L, Underwood G, Chapman P. Enhanced memory for emotional pictures: A product of increased attention to affective stimuli? European Journal of Cognitive Psychology 2010;22(8):1235-1247.

[92] Eichenbaum H. Prefrontal-hippocampal interactions in episodic memory. Nature Reviews Neuroscience 2017;18(9):547-558.

[93] Robin J, Hirshhorn M, Rosenbaum RS, Winocur G, Moscovitch M, Grady CL. Functional connectivity of hippocampal and prefrontal networks during episodic and spatial memory based on real-world environments. Hippocampus 2015;25(1):81-93.

[94] Jin J, Maren S. Prefrontal-hippocampal interactions in memory and emotion. Frontiers in Systems Neuroscience 2015;9:170-170.

[95] Sampath D, Sathyanesan M, Newton SS. Cognitive dysfunction in major depression and Alzheimer's disease is associated with hippocampus-prefrontal cortex dysconnectivity. Neuropsychiatric Disease and Treatment 2017;13:1509-1519.

[96] Emrich SM, Riggall AC, LaRocque JJ, Postle BR. Distributed patterns of activity in sensory cortex reflect the precision of multiple items maintained in visual short-term memory. Journal of Neuroscience 2013;33(15):6516-6523.

[97] Wheeler ME, Petersen SE, Buckner RL. Memory's echo: vivid remembering reactivates sensory-specific cortex. Proceedings of the National Academy of Sciences 2000;97(20):11125-11129.

[98] Chen Y, Spagna A, Wu T, Kim TH, Wu Q, Chen C, et al. Testing a cognitive control model of human intelligence. Scientific Reports 2019;9(1):1-17.

[99] Badre D, Wagner AD. Selection, integration, and conflict monitoring: assessing the nature and generality of prefrontal cognitive control mechanisms. Neuron 2004;41(3):473-487.

[100] McGuire JT, Botvinick MM. Prefrontal cortex, cognitive control, and the registration of decision costs. Proceedings of the National Academy of Sciences 2010;107(17):7922-7926.

[101] Cole MW, Yarkoni T, Repovs G, Anticevic A, Braver TS. Global connectivity of prefrontal cortex predicts cognitive control and intelligence. Journal of Neuroscience 2012;32(26):8988-8999.

[102] Koechlin E, Ody C, Kouneiher F. The architecture of cognitive control in the human prefrontal cortex. Science 2003;302(5648):1181-1185.

[103] Chudasama Y. Animal models of prefrontal-executive function. Behavioral Neuroscience 2011;125(3):327.

[104] Moore TL, Schettler SP, Killiany RJ, Rosene DL, Moss MB. Effects on executive function following damage to the prefrontal cortex in the rhesus monkey (Macaca mulatta). Behavioral Neuroscience 2009;123(2):231.

[105] Badre D, Hoffman J, Cooney JW, D'Esposito M. Hierarchical cognitive control deficits following damage to the human frontal lobe. Nature Neuroscience 2009;12(4):515-522.

[106] Lesh TA, Niendam TA, Minzenberg MJ, Carter CS. Cognitive control deficits in schizophrenia: mechanisms and meaning. Neuropsychopharmacology 2011;36(1):316-338.

[107] Gazzaley A, Rissman J, Cooney J, Rutman A, Seibert T, Clapp W, et al. Functional interactions between prefrontal and visual association cortex contribute to top-down modulation of visual processing. Cerebral Cortex 2007;17(suppl_1):i125-i135.

[108] Zanto TP, Rubens MT, Thangavel A, Gazzaley A. Causal role of the prefrontal cortex in top-down modulation of visual processing and working memory. Nature Neuroscience 2011;14(5):656-661.

[109] Alvarez JA, Emory E. Executive function and the frontal lobes: a meta-analytic review. Neuropsychology Review 2006;16(1):17-42.

[110] Chrysikou EG, Weber MJ, Thompson-Schill SL. A matched filter hypothesis for cognitive control. Neuropsychologia 2014;62:341-355.

[111] Diamond A. Executive functions. Annual Review of Psychology 2013;64:135.

[112] Cole MW, Schneider W. The cognitive control network: integrated cortical regions with dissociable functions. Neuroimage 2007;37(1):343-360.

[113] Chouinard PA, Paus TÅ. The primary motor and premotor areas of the human cerebral cortex. The Neuroscientist 2006;12(2):143-152.

[114] Koch G, Franca M, Del Olmo MF, Cheeran B, Milton R, Sauco MA, et al. Time course of functional connectivity between dorsal premotor and contralateral motor cortex during movement selection. Journal of Neuroscience 2006;26(28):7452-7459.

[115] McDowell JE, Dyckman KA, Austin BP, Clementz BA. Neurophysiology and neuroanatomy of reflexive and volitional saccades: evidence from studies of humans. Brain & Cognition 2008;68(3):255-70.

[116] Vernet M, Quentin R, Chanes L, Mitsumasu A, Valero-Cabré A. Frontal eye field, where art thou? Anatomy, function, and non-invasive manipulation of frontal regions involved in eye movements and associated cognitive operations. Frontiers in Integrative Neuroscience 2014;8:66.

[117] Burman DD, Lie-Nemeth T, Brandfonbrener AG, Parisi T, Meyer JR. Altered finger representations in sensorimotor cortex of musicians with focal dystonia: Precentral cortex. Brain Imaging and Behavior 2009;3(1):10-23.

[118] Rose M, Haider H, Salari N, Buchel C. Functional dissociation of hippocampal mechanism during implicit learning based on the domain of associations. Journal of Neuroscience 2011;31:13739-13745.

[119] Cole MW, Bassett DS, Power JD, Braver TS, Petersen SE. Intrinsic and task-evoked network architectures of the human brain. Neuron 2014;83(1):238-251.

[120] Brooks SJ, Savov V, Allzen E, Benedict C, Fredriksson R, Schioth HB. Exposure to subliminal arousing stimuli induces robust activation in the amygdala, hippocampus, anterior cingulate, insular cortex and primary visual cortex: A systematic meta-analysis of fMRI studies. NeuroImage 2012;59(3):2962-2973.

[121] Degonda N, Mondadori CR, Bosshardt S, Schmidt CF, Boesiger P, Nitsch RM, et al. Implicit associative learning engages the hippocampus and interacts with explicit associative learning. Neuron 2005;46(3):505-20.

[122] Schendan HE, Searl MM, Melrose RJ, Stern CE. An FMRI study of the role of the medial temporal lobe in implicit and explicit sequence learning. Neuron 2003;37(6):1013-1025.

[123] Foerde K, Knowlton BJ, Poldrack RA. Modulation of competing memory systems by distraction. Proceedings of the National Academy of Sciences 2006;103(31):11778-83.

[124] Squire LR, Dede AJO. Conscious and unconscious memory systems. Cold Spring Harbor Perspectives in Biology 2015.

[125] Ashby FG, Turner BO, Horvitz JC. Cortical and basal ganglia contributions to habit learning and automaticity. Trends in Cognitive Sciences 2010;14(5):208-215.

Section 2

Disease and Treatment

Chapter 5

Tau in Health and Neurodegenerative Diseases

Dandan Chu and Fei Liu

Abstract

Tau, one of the major microtubule-associated proteins, modulates the dynamic properties of microtubules in the mammalian nervous system. Tau is abundantly expressed in the brain, particularly in the hippocampus. Insoluble and filamentous inclusions of tau in neurons or glia are discovered in neurodegenerative diseases termed 'tauopathies', including Alzheimer's disease (AD), argyrophilic grain disease (AGD), corticobasal degeneration (CBD), frontotemporal dementia (FTD), Pick's disease (PiD) and progressive supranuclear palsy (PSP). Accumulation of intracellular neurofibrillary tangles (NFTs), which are composed of hyperphosphorylated tau, is directly correlated with the degree of Alzheimer's dementia. This chapter reviews the role of tau protein in physiological conditions and the pathological changes of tau related to neurodegenerative diseases. The applications of tau as a therapeutic target are also discussed.

Keywords: tau, neurodegenerative diseases, pathology, Alzheimer's disease, therapy

1. Introduction

The human hippocampus is critically involved in spatial navigation, the forming, storing and retrieval of episodic memory and the processing of many other types of memory [1]. In the normal human brains, the hippopotamus volume remains relatively stable before the age of 60, and then gradually atrophies. In Alzheimer's disease (AD) patients, the pattern of brain atrophy follows a stereotypical pathway that initiates from the entorhinal cortex and the hippocampus, and then spreads to the medial parietal, lateral temporal and frontal regions, eventually to the neocortex. As the connections between the hippocampus and its neighboring cortical structures are selectively vulnerable to neurodegeneration in AD, hippocampal volume loss is considered an important indicator of AD neuropathology [2]. In addition, since neurogenesis in adult brain only occurs in the dentate gyrus of the hippocampus and the olfactory bulb, hippocampus atrophy in AD also alters the production of newborn neurons [3].

Tau, initially isolated as a microtubule-associated protein from the porcine brain in 1975, is predominantly expressed in the hippocampus [4, 5]. In the previous year, neurofibrillary tangles (NFTs) and a paired helical filament (PHF) protein had been identified from the brains of patients with Alzheimer's disease (AD) [6], but it was not until 1986 that tau was discovered to be a major component of PHF [7]. Subsequently, tau was implicated in the pathogenesis of over 25 human neurological disorders (termed "tauopathies"), including Alzheimer's disease (AD), argyrophilic grain disease (AGD), corticobasal degeneration (CBD), frontotemporal dementia

(FTD), globular glial tauopathy (GGT), primary age-related tauopathy (PART), Pick's disease (PiD) and progressive supranuclear palsy (PSP) [8–10].

As the most common neurodegenerative disease, AD is characterized by early impairments in learning and memory, followed by progressive loss of complex attention, executive function, language, orientation and self-care ability, changes in mood, loss of motivation, and impairments in thinking, behavior and/or social comportment [11]. The two major neuropathological hallmarks of AD are the extracellular deposition of β-amyloid (Aβ) plaques and the intracellular neurofibrillary tangles (NFTs) consisting of aggregated hyperphosphorylated tau [12]. NFTs pathology in AD is initiated in the locus coeruleus and transentorhinal cortex, from where it spreads to the limbic system (e.g., entorhinal cortex and hippocampus) and further to the neocortex, leading to six Braak stages [13]. The progression of cognitive decline in AD correlates with the accumulation of NFTs and loss of hippocampal volume, but not deposition of Aβ plaques [14, 15]. Since most therapies targeting Aβ failed in late-stage clinical trials for AD in the past decades, increasing research revealing the roles of tau in disease has inspired tau-targeting approaches in the treatment of AD and related tauopathies [16, 17].

This chapter reviews the expression and functions of tau in physiological conditions, the pathological changes of tau in diseases, such as genetic variants, post-translational modifications (PTMs) and prion-like seeding and propagation. Recent advances in the development of tau-based therapies for AD and other neurodegenerative diseases are also discussed.

2. Tau gene

Human tau protein is encoded by the microtubule-associated protein tau (MAPT) gene, which locates on chromosome 17q21.31 and consists of 16 exons. In the central nervous system (CNS), the alternative splicing of exons 2, 3, and 10 gives rise to six tau isoforms with zero (0 N), one (1 N) or two (2 N) N-terminal inserts and three (3R-tau) or four (4R-tau) microtubule-binding repeats (**Figure 1**) [18]. The longest isoform of human brain tau consists of 441 amino acids (2N4R, tau441) with an apparent molecular weight (MW) of 46 kDa. Exons 4a and 6 are predominantly expressed in the peripheral nervous system (PNS), producing proteins of apparent MW of 110 kDa, named big tau [19].

2.1 Alternative splicing of tau pre-mRNA

The expression of tau isoforms is developmentally and pathologically regulated. 3R-tau isoforms are expressed throughout life, including in the fetal brain, whereas 4R-tau isoforms are specifically expressed in adults, resulting in approximately equal levels of 3R-tau and 4R-tau in the adult human brain [18]. Rodent tau shares about 90% homology with human tau. Unlike humans, rodents express 3R-tau only in fetus and infant, and mainly 4R-tau in adulthood [20].

Tau pre-mRNA contains multiple cis-elements that allow the interaction of trans-acting factors like the serine and arginine-rich (SR) proteins and heterogeneous nuclear ribonucleoproteins (hnRNPs), including an SC35-like enhancer, a polypurine enhancer (PPE) and an A/C-rich enhancer (ACE) at the 5′ terminus, and an exonic splicing silencer (ESS) and an exonic splicing enhancer (ESE) at the 3′ end of tau exon 10, and an intronic splicing silencer (ISS) and an intronic splicing modulator (ISM) at 5′ end of intron 10 [21]. Binding of trans-acting factors to these cis-elements either suppresses (SRSF3, SRSF4, SRSF7, SRSF11, U2AF, PTB

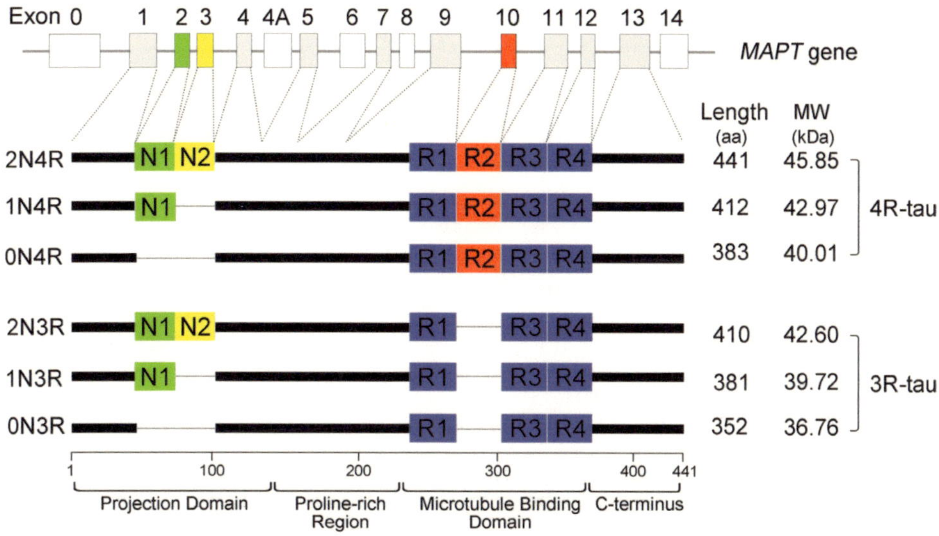

Figure 1.
Gene structure and expression of human MAPT gene. MAPT gene is localized to chromosome 17 and consists of 16 exons. Exons 4a, 6, and 8 are expressed in PNS. Alternative splicing of tau exon 2 (green, encodes N1) or exon 3 (yellow, encodes N2) produces isoforms with zero (0 N), one (1 N) or two (2 N) N-terminal inserts. Exons 9, 11 and 12 encode the microtubule-binding repeats R1, R3, and R4 (blue). Alternative splicing of exon 10 (red, encodes R2) generates isoforms containing four (4R-tau) or three (3R-tau) microtubule-binding repeats.

and hnRNP G) or promotes (hTRA2-beta1, CELF3, CELF4, SRSF1, SRSF2, SRSR6, SRSF9, RNA helicase p68, RNA binding motif protein 4 and Tar DNA-binding protein 43) the inclusion of exon 10 [22].

Mutations in the MAPT gene or dysregulation of the splicing factors that alter the expression of tau exon 10 are involved in the pathogenesis of neurodegenerative diseases [21]. PiD is a prototypical 3R-tauopathy with spherical tau inclusions (Pick bodies), whereas PSP, CBD, AGD and GGT are primary 4R-tauopathies. AD, FTD and chronic traumatic encephalopathy (CTE), progressive neurodegeneration associated with repetitive mild traumatic injury, comprise both 3R and 4R-tau pathologies [23]. Studies on the amount of the six tau transcripts in AD brain have been contradictory [24, 25]. However, 4R- and 3R-tau proteins appear equally in the early-stage and some late-stage AD cases [26, 27]. In some cases of late-stage AD, 3R-tau protein is predominantly expressed in the subiculum, entorhinal cortex and area CA1 of the hippocampus [27], implying that dysregulation of tau exon 10 splicings could be related to AD progression rather than initiation.

2.2 Tau mutations

So far, 112 mutations have been identified in human the MAPT gene, of which 13 were located in an intron (https://www.alzforum.org/mutations). Not all the mutations are pathogenic, and at least 27 benign mutations are not responsible for significant clinical symptoms. The largest number of MAPT mutations (55) is found in FTD. Second, 15 mutations are identified in AD, most of which are benign mutations causing no significant neuropathology. The rest mutations are correlated to PSP (10 mutations), Parkinson's disease (PD) (8), PiD (7), CBD (4), AGD (1) and other tauopathies (**Table 1**). As genetic risk factors for neurodegenerative diseases, pathogenic tau mutations alter the protein sequence or/and the balance between 4R- and 3R-tau by changing alternative splicing [28].

Clinical Phenotype	Mutations
Argyrophilic grain disease	S305I
Alzheimer's Disease	A90V, A152T, G213R, V224G, Q230R, K280del, V287I, A297V, S318L, R406W, L410F*, S427F*, P512H* Duplication 17q21.31, IVS10 + 16 C > T
Corticobasal Degeneration	A152T, C291R, P301T, N410H
Frontotemporal Dementia	R5H, G55R, V75A, A152T, A239T, D252V, I260V, L266V, G272V, G273R, N279K, K280del, L284L, N296D, N296N, N296H, P301L, P301P, P301S, P301T, S305N, S305S, L315L, L315R, S320F, P332S, G335A, G335S, G335V, Q336R, V337M, E342V, S356T, V363I, P364S, G366R, K369I, E372G, G389_I392del, G389R, G389R, P397S, R406W, T427M IVS9–10 G > T, IVS9–15 T > C, IVS10 + 3 G > A, IVS10 + 4 A > C, IVS10 + 11 T > C, IVS10 + 12 C > T, IVS10 + 13 A > G, IVS10 + 14 C > T, IVS10 + 15 A > C, IVS10 + 19 C > G, IVS10 + 16 C > T
Pick's disease	G272V, K280del, S320F, Q336R, Q336H, K369I, G389R
Progressive Supranuclear Palsy	R5L, A152T, L284R, S285R, N296N, N296del, G303V, S305S, D285N* IVS10 + 16 C > T
Parkinson's Disease	R5C, A41T, A152T, N296del, I360V, S427F*, T427M, R448*

Table 1.
MAPT mutations found in main neurodegenerative diseases.

**The position of this variant is about the longest isoform of peripheral tau, which is 776 amino acids in length.*

Two major extended haplotypes cover the MAPT gene: H1 and H2 [29]. The frequency of the haplotypes differs between population groups; the H2 haplotype is found primarily in Caucasian and southwest Asians, but barely reported in the Chinese Han population [29, 30]. H1 haplotype was shown overrepresented in Caucasian patients with PSP [31]. Nevertheless, the relationship between MAPT haplotype and AD is still contradictory [32].

3. Tau protein

3.1 Protein structure of tau

Tau is a highly water-soluble and basic protein with little secondary structure. As an intrinsically disordered protein, tau consists of a large number of serine/threonine and arginine/lysine/histidine residues, which makes the protein easy to be hyperphosphorylated. Full-length human tau (tau441) is composed of four domains: (i) an acidic amino-terminal projection domain that projects away from the surface of the microtubule, (ii) a proline-rich region required for the interaction with SH3-domain-containing proteins like tau kinases, (iii) a microtubule-binding domain involved in mediating tubulin assembly and tau aggregation, and (iv) a C-terminus (**Figure 1**) [33].

3.2 Localization of tau

Human tau is highly expressed in the frontal and temporal cortices and is decreased to approximately a quarter in the cerebellum. In rodent brains, the highest levels of tau are detected in the hippocampus and entorhinal cortex. The cerebellum and olfactory bulb showed the lowest total tau level, being about 2/3 of that in the frontal cortex [5]. The expression of 3R-tau is comparable throughout different regions in the adult rat brains, such as the hippocampus, entorhinal cortex, frontal cortex, occipital-temporal cortex, parietal–temporal cortex, striatum, thalamus, olfactory bulb and cerebellum. However, the distribution of 4R-tau showed significant regional differences, with the highest levels in the entorhinal and frontal cortices and the lowest in the cerebellum. The uneven expression of tau protein in brain regions may contribute to distinct vulnerability/resistance to tau pathology [5].

In the human brain, tau is primarily expressed in neurons and also expressed at lower levels in oligodendrocytes and astrocytes [34]. In physiological conditions, tau was believed to be predominantly localized to axons, but limited in the soma and dendrites of neurons. Tau levels are comparable in gray matter and white matter in normal elderly brains, but higher in gray matter than whiter matter in AD brains [35]. Normally, tau monomer is difficult to be immunostained. Only aggregated tau or microtubule-binding tau can be detected by immunostaining. Although high levels in axons, a pre-synaptic abundance of tau is low. The mechanism for polarized neuronal distribution of tau could be (i) relocalization of tau from axon to soma may be blocked by the axon initial segment (AIS), (ii) Annexin A2 in the distal part of the axon interacts with tau and provides a sink for the redistribution of tau [16]. Under pathological conditions, endogenous tau translocates from axon to the soma and dendrite, and into the post-synapse. Early studies in AD and FTD have revealed that NFTs composed of aggregated hyperphosphorylated tau are localized in the soma and dendrites [36]. Translocation of tau depends on its interaction with microtubules [37] and are modulated through multiple mechanisms involving pathological post-translational modification (e.g., hyperphosphorylation) and/or non-physiological overexpression of tau protein [38], imbalanced expression of tau isoforms (e.g., 2 N tau) [39] and dysregulation of extracellular signals [16]. Furthermore, tau messenger RNA (mRNA) might be recruited to the dendrite and post-synapses, resulting in local somatodendritic translation of tau protein [40, 41]. Lastly, tau is also detected in the nucleus, where it is likely to protect DNA integrity from stress [42].

3.3 Function of tau

Microtubules are more stable in mature neurons than in non-neuronal cells [43], probably due to posttranslational modifications of the tubulin subunits as well as their interaction with specific neuronal microtubule-associated proteins (MAP) such as tau [44]. Tau binds to the interface between tubulin heterodimers through its microtubule-binding repeats [45] and stabilizes microtubules in the test tube and cultured cells [46, 47]. 4R-tau isoforms have a stronger affinity for microtubules than 3R-tau isoforms [48] and are more prone to promote microtubule assembly [49]. Besides, tau can nucleate and bundle microtubules in vitro [50] as well as in axons of mammalian neurons [51]. Recent studies have shown that tau not only acts as a microtubule stabilizer but also positively regulates the elongation of labile domains of microtubules at the plus ends [43]. Additionally, tau is involved in mediating intracellular transport along the axon [52], synaptic structure and function, and signaling pathways in neurons [53].

In AD and other tauopathies, pathological tau would detach from microtubules, leading to decreased microtubule stability, impaired axonal transport and synaptic function [54]. First, tau is more abundant on the labile domains of microtubules to protect them from severing proteins like katanin. Under a pathological condition, hyperphosphorylated tau disassociates with the microtubules and causes rapid and selective shrinkage of microtubule labile domains [55]. Disease-associated tau mutants, like K369I, G389R found in FTD, also showed decreased associations rate with microtubules, resulting in reduced ability in promoting microtubule assembly

[56]. Second, the ability of tau to regulate axonal transport alters under pathological conditions. Amino acids (aa) 2–18 of tau protein, termed the phosphatase-activating domain (PAD), activate a signaling cascade involving protein phosphatase 1 (PP1) and glycogen synthase kinase 3β (GSK-3β) that results in disruption of kinesin-1-mediated anterograde fast axonal transport. Y18 (tyrosine 18) phosphorylation of tau, which is stronger in monomers than in filaments, shows reduced inhibition of kinesin-1 and is significantly reduced in disease-associated tau species [57]. Moreover, intracellular tau aggregates have been shown to impair fast axonal transport by increasing the run length, run time and instantaneous velocity of membranous organelles [58]. Third, pathological tau accumulates on both pre- and post-synapses in the AD brain. Presynaptic accumulation of tau induces the depletion of the synaptic vesicle pool, followed by impaired synaptic transmission and plasticity [59]. On the other hand, toxic Aβ oligomers trigger N-methyl-D-aspartate receptor (NMDAR)-mediated excitotoxicity depending on the presence of endogenous tau [16].

Besides, glial tau pathology is also a common feature of many tauopathies and contributes to pathogenesis [34]. Oligodendrocytic tau pathology disrupts the maintenance of myelin sheath [60]. Although found at trace levels in astrocytes [61], tau has been shown to accumulate in astrocytic end feet directly apposed to vascular endothelial cells, and therefore impair the blood–brain barrier (BBB) integrity [62] and cause neuron degeneration without neuronal tau inclusions [63]. Filamentous recombinant tau also activates astrocytes via integrin signaling [64]. The expression and function of tau in microglia remain unclear [34]. However, microglia may regulate the uptake and exosomal secretion of tau, therefore involved in the spreading of tau pathology across the brain [65].

4. Post-translational modifications (PTMs) of tau

Tau is post-translationally modified by multiple mechanisms, including phosphorylation, ubiquitination, acetylation, methylation, glycosylation, glycation, nitration, lipoperoxidation, sumoylation and truncation [22]. Recently, up to 95 PTMs have been identified in human brains by high-resolution quantitative proteomics (**Table 2**) [70]. In neurodegenerative diseases, tau undergoes a series of pathological changes, such as PTMs alterations and the prion-like seeding and propagation, many of which not just accompany the diseases and indicate the pathology progression and individual heterogeneity, but are also the driving force of diseases.

4.1 Phosphorylation

Tau is a phosphoprotein. Theoretically, 80 serine/threonine and 5 tyrosine residues of the longest CNS tau isoform-tau441 can potentially be phosphorylated. In the brains of normal elderly, tau contains 2–3 mol phosphate/mole of the protein [71]. Phosphorylation sites like pT231, pT181, pS404 are found with high frequency (> 50%) [70]. However, tau phosphorylation increases 2–3 times in AD brains, i.e. hyperphosphorylation [71]. At least 55 phosphorylation sites are detected on pathological tau (**Table 2**) [70]. Hyperphosphorylation of soluble, oligomeric and seed-competent tau all exhibit substantial patient-to-patient heterogeneity, even though certain hot spots are present in AD brains, including pT181, pT217, pT231/S235, pS262, pS396 and pS400/T403/S404 [70, 72]. Among these sites, the phosphorylation levels of pT231/S235 and pS262 are positively correlated to the seeding capacity of tau species [72].

Rodent tau exhibits approximately 90% homology with human tau. Profiling of PTMs showed modifications on up to 63 sites on tau protein in the wild-type mouse brain, of which 27 were phosphorylation sites [73]. Phosphorylation of tau at specific

Modification	Sites
Phosphorylation	Y29, T30, T39, S46, S56, S68, T69, T71, T102, T111, S113, T153, S185, S184, T181, T175, S191, S198, S199, S202, T205, S210, T212, S214, T217, T220, T231, S235, S237, S238, S241, S258, S262, T263, S289, S293, S305, Y310, S316, S352, S356, T361, T386, Y394, S396, S400, T403, S404, S409, S412, S413, T414, S422, S433, S435
Ubiquitination	K163, K180, K190, K224, K228, K234, K240, K254, K257, K259, K267, K274, K280, K281, K290, K298, K311, K317, K321, K331, K340, K343, K353, K369, K375, K383, K385, K395
Acetylation	K24, K44, K240, K267, K274, K280, K281, K298, K311, K317, K331, K343, K347, K353, K375, K370, K369, K385, K395
Methylation	K44, K67, K87, K163, K174, K180, K254, K267, K290, R406, K438
SUMOylation	K340
Nitration	Y18, Y29, Y197, Y394
Truncation	M1-A2, M11-E12, V10-M11, D13-H14, D25-Q26, K44-E45, T102-A103, T123-Q124, R126-M127, A152-T153, R155-G156, I171-P172, A173-K174, G186-E187, Y197-S198, P223-K224, R230-T231, S237-S238, A239-K240, R242-L243, N255-V256, S258-K259, I260-G261, N279-K280, K281-L282, S305-V306, Q307-I308, I308-V309, Y310-K311, D314-L315, H330-K331, K340-S341, N368-K369, A390-E391, E391-I392, Y394-K395, D402-T403, D421-S422

Table 2.
Tau PMTs found in AD brains [50, 66–69].

sites was different in the brains of human and AD model mice. For example, 3 × TG-AD mice, the commonly-used AD model which contains three mutations associated with familial AD (APP Swedish, MAPT P301L, and PSEN1 M146V), are not significantly hyperphosphorylated at pS199, pS214, pS396/S400 and pS422 as in AD brains, implying a limitation of a mouse model in studying human tau pathology [74].

Abnormal hyperphosphorylation of tau is a pivotal step in neurofibrillary degeneration in AD and other tauopathies [71]. Taking the AD brain as an example, tau can be separated into three pools according to the phosphorylation state and solubility: (i) non-hyperphosphorylated normal tau (AD-tau), (ii) hyperphosphorylated tau (AD P-tau) and (iii) polyubiquitinated, hyperphosphorylated and aggregated tau in the insoluble PHFs (PHF-tau) [75]. Hyperphosphorylation of tau induces pathology through multiple mechanisms. First, hyperphosphorylation reduces tau affinity for microtubules. Natural tau forms a "paper clip" structure, with the N- and C-terminus fold over the microtubule-binding domain to prevent self-aggregation [76]. Hyperphosphorylation changes the net charge of tau protein and alters tau conformation to expose the microtubule-binding domain, thereby facilitating self-oligomerization and aggregation. Hyperphosphorylated and/or aggregated tau detach from the microtubules and lose their ability to stabilize microtubules [71]. Besides normal, AD P-tau captures microtubule-associated proteins other than tau, such as MAP1 and MAP2 [77], leading to further disruption of microtubules. Second, hyperphosphorylated tau redistributes from axons to the somatodendritic compartment and impairs synaptic function (see Section 3.3) [16, 54]. Third, phosphorylation may change the interaction of tau with other regulatory proteins [28]. Recently, 75 proteins specifically bound to phosphorylated tau in NFTs have been identified by quantitative proteomics coupled with affinity purification-mass spectrometry; most enriches in the protein ubiquitination pathway and phagosome maturation [78]. Whether hyperphosphorylation of tau alters its affinity to these proteins and directly leads to the damage of relevant pathways deserves more extensive investigation.

Tau phosphorylation is regulated by both proline-directed [GSK-3β, cyclin-dependent-like kinase-5 (CDK5), dual-specificity tyrosine phosphorylation regulated kinase 1A (DYRK1A) and extracellular signal-related protein kinase (Erk)] and non-proline-directed protein kinases [calcium/calmodulin activated protein kinase II (CaMKII), protein kinase A (PKA), casein kinase 1 (CK1) and microtubule affinity-regulated kinase 110 (MARK p110)] in vivo, which are activated and/ or overexpressed in AD brains [22]. On the contrary, tau is dephosphorylated by protein phosphatases, in particular protein phosphatases 2A (PP2A), which is responsible for over 70% of the total tau phosphatase activity in the human brain [79]. Dephosphorylation of AD P-tau with PP2A restores tau activity in promoting microtubule assembly in vitro and diminishes AD P-tau-induced propagation of tau pathology in mouse brain [80]. In disease conditions, phosphatase activity against tau is reduced to half, further increasing the imbalance between kinase and phosphatase activities, eventually resulting in excessive phosphorylation of tau [79]. It should be noted that tau is rapidly dephosphorylated during postmortem in a site-specific manner, suggesting timely dissection and proper cooling of the brain tissues [81].

4.2 Ubiquitination

Ubiquitination, a PTM that covalently conjugated ubiquitin (a highly conserved 76 amino acid protein) to the ε-amino group of target lysine residues in a protein, is usually involved in cellular protein degradation as well as non-degradative pathways including cell signaling, mitochondrial homoeostasis and DNA damage responses [82]. Ubiquitin positive pathological aggregates are present in AD, FTD, PD, CBD and other neurodegenerative diseases [83]. Quantitative analysis of ubiquitylome in AD brain reveals 28 ubiquitination sites in tau protein, which are the most abundant PTMs except phosphorylation (**Table 2**) [83]. Most of these ubiquitylation sites are located in the proline-rich region and the microtubule-binding domain.

Ubiquitination of tau is catalyzed by various ubiquitin ligases (E3 ligases), for instance, the C-terminus of the Hsc70-interacting protein (CHIP), TNF receptor-associated factor 6 (TRAF6) and axotrophin/MARCH7 [22]. High molecular weight (HMW) tau extracted from AD brain is shown to be polyubiquitinated, likely through K6-, K11- or K48-linkages [83], while PHF-tau is mostly monoubiquitylated, making it insufficient to trigger the ubiquitin-proteasome system (UPS)-mediated proteolysis [84]. Both polyubiquitylation and monoubiquitylation of tau contribute to the formation of insoluble protein inclusions [85, 86].

As a natively unfolded protein, tau is degraded by ATP/ubiquitin-independent 20S proteasome in physiological conditions [87]. Misfolded tau is typically ubiquitylated and is sent to the proteasome for degradation [88]. However, misfolded oligomers and aggregates cannot be fully degraded by the proteasome, but also directly damage the proteasome activity [89]. The autophagy-lysosome system provides a more potent pathway to degrade tau aggregates, which also relies on ubiquitination modification for recognition [88]. Therefore, activating ubiquitin degradation for toxic tau species is considered as one of the potential therapeutic strategies for the treatment of AD and related tauopathies.

4.3 Acetylation

Acetylation is a co- or post-translational modification that utilizes acetyl-CoA as the acetyl source to modify the N-termini or specific lysine residues in proteins [84]. Tau acetylation is catalyzed by the histone acetyltransferase p300 (EP300) or CREB-binding protein (CBP), and removed by sirtuin 1 (SIRT1) and histone deacetylase 6 (HDAC6) [22]. Tau is also able to catalyze self-acetylation by using cysteine residues C291 and C322 in the R2 and R3 repeats, respectively. 4R-tau displays the higher activity of autoacetylation than 3R-tau because the latter lacks the R2 repeat [90]. 19 distinct acetylation sites have been mapped in tau protein isolated from AD brains, most of which are located within the microtubule-binding repeats and the flanking region (**Table 2**) [70].

The pathological effect of tau acetylation depends on specific modification sites. For example, high levels of tau acetylation are found at Lys163, Lys174, Lys180, Lys274, Lys280, Lys281 and Lys369 in AD brains, which may be related to the impairment of tau function [84]. Acetylation at these sites could prevent the polyubiquitylation and degradation of hyperphosphorylated tau, thus accelerating the accumulation of phosphorylated tau and promoting NFTs formation, accompanied by increased cognitive impairment. Acetylated tau also mislocalizes to the somatodendritic compartment and disrupts cytoskeleton dynamics, postsynaptic protein localization and receptor trafficking, consequently giving rise to synaptic plasticity deficits and memory loss [91]. Moreover, auto-acetylation of tau in C291 and C322 is coupled to its auto-proteolysis at K281-L282 and K340-S341 [90]. On the contrary, Lys259, Lys290, Lys321 and Lys353 within the KXGS motifs, which are found hypoacetylated in the AD brain, are normally acetylated to inhibit tau phosphorylation and aggregation [92].

4.4 Truncation

As an intrinsically disordered protein, tau is sensitive to proteolysis. In addition to acetylation-induced auto-proteolysis, tau can be cleaved by a variety of proteases both in vitro and in vivo including a disintegrin and metallopeptidase domain 10 (ADAM-10), asparagine endopeptidase (AEP), Calpain-1, Calpain-2, Caspase-3, Caspase-6, chymotrypsin, thrombin, cathepsins, human high-temperature requirement serine protease A1 (HtrA1) and puromycin-sensitive aminopeptidase (PSA). Besides, many fragments of tau in the brain have been found with undefined proteases (**Table 2**) [74, 93, 94]. In NFTs, at least three site-specific cleavages of tau (N368, E391 and D421) have been identified to be correlated with the progression of Braak stages [95]. Similar cleavage sites are also detected in several mouse models of tauopathies [28]. To date, more than 50 truncated forms of tau have been identified and over 30 are present in AD brains [94].

Truncation of tau plays a crucial role in its pathology. Truncation opens up the “paper clip” tertiary structure of tau protein, increases its site-specific phosphorylation, self-aggregation and affinity to oligomeric tau derived from AD brain (AD O-Tau), and thus promotes tau capture and seeded-aggregation by AD O-Tau [28, 96]. Tau truncation alone is sufficient to trigger hyperphosphorylation and aggregation [97]. Some fragments of tau can spread transcellularly, resulting in the propagation of tau pathology [94]. Additionally, Tau truncation can also induce toxic responses like apoptosis which is independent of its function on aggregates [98].

The characteristics of tau fragments depend on their amino acid composition. The C-terminal truncation increased tau fibrillization in vitro [66], while N-terminal truncations are found more associated with hyperphosphorylated high molecular weight tau oligomers (HMW-tau) isolated from AD brains [74, 93]. Tau fragments containing the aggregation-prone elements (**Table 3**) are prone to assemble the protease-resistant pathological core, which has various compositions in the different tauopathies [99]. Some specific fragments are secreted from the cytosol into the interstitial fluid and further released to the cerebrospinal fluid (CSF) or plasma, making them potentially biomarkers to indicate the progression of AD and other tauopathies [100]. Early study using immunoprecipitation showed that tau in CSF is predominantly the N-terminus fragments with an apparent molecular weight of approximately 20–40 kDa [101]. A 20–22 kDa NH_2-truncated form of tau (aa 26–230) identified in CSF is found to be neurotoxic due to its damages to mitochondrial oxidative phosphorylation [102].

Disease	Aggregation-prone elements
AD	3R Q273-E380, 4R-tau: G304-E380
CBD	K274-E380
CTE	3R-tau: K274-R379, 4R-tau: S305-R379
PiD	K254-F378

Table 3.
Aggregation-prone elements of tau in representative tauopathies [5, 188].

High-resolution mass spectrometry revealed at least 19 tau fragments in the CSF, of which tau aa 156–224 is the most abundant peptide [67, 68]. Nonetheless, the prion-like seeding activity of CSF tau fragments is quite limited [69].

4.5 Other PMTs

In addition to the above modifications, tau can also be modulated by methylation, SUMOylation, nitration, glycosylation and glycation. The contributions of these PMTs to tau pathology are mostly unelucidated.

Tau methylation mainly occurs on lysine residues in the proline-rich region and the microtubule-binding domain, and a few arginine residues [73]. Up to 11 methylation sites were found in the human brain [70]. Methylated tau is highly concentrated in NFTs particularly in late-stage AD brains [103]. Lysine methylation suppresses tau binding to the microtubule-binding domain, increases abnormal phosphorylation of tau and blocks the UPS-mediated tau degradation. However, the role of methylation on tau self-aggregation is still controversial [84].

K340 residue is the major site of tau that be modified by the small ubiquitin-like modifier protein (SUMO) [104]. SUMO-1 colocalizes with phosphorylated tau in the AD brain. Sumoylation reciprocally stimulates tau phosphorylation at T231 and S262, and competes against K340 ubiquitylation and consequently suppresses degradation [105].

Tau can be nitrated on four Tyr residues, Y18, Y29, Y197 and Y394, which are found in AD and non-AD tauopathies [106]. The effect of nitration on tau assembly depends on the specific nitration sites in vitro. Y18 nitration is reported to be associated with astrocyte activation [107].

In addition, the complexity of PTM is reflected more in the cross-talk between various modifications, not just in the types of modifications. First, a single amino acid residue could be modified by different PMTs. Taking lysine residues as an example, some of these residues in tau protein are competitively modified by ubiquitylation, methylation, acetylation, or SUMOlyation in the AD brain (**Table 2**). The competition between these PMTs will determine whether tau will undergo degradation or pathological transformation [84]. Secondly, PMTs at different sites cross-talk with each other. Many phosphorylation and ubiquitination sites of tau are located within the KXGS motifs in the microtubule-binding domain. Tau hyperphosphorylation is shown to facilitate ubiquitylation of NFT tau [83]. Therefore, further investigations focusing on the cross-talk between tau PMTs are required to reveal and intervene in the pathological changes of tau.

5. Propagation of tau aggregation

In AD brains, the progression of tau aggregation follows a stereotypical pattern of spread (the Braak stages): initiates from the locus coeruleus and transentorhinal cortex

(Braak stages I and II), gradually spread to the limbic system (Braak stages III and IV) and eventually to the neocortex (Braak stages V and VI) [108]. The stereotypical transmission of tau pathology is highly correlated with the cognitive impairment in AD [13]. Remarkably, tau pathology can be induced in rodent brains by injecting aggregates isolated from AD brains, and propagating to anatomically connected brain regions, in a similar spreading pattern to that observed in AD patients [80]. Besides, injection of tau aggregates extracted from other neurodegenerative diseases, such as AGD, CBD and PSP, also recapitulated the hallmark lesions of corresponding diseases [109]. A large amount of evidence indicates that the prion-like propagation of misfolded tau may explain the diverse characteristics of tauopathies [10].

5.1 Tau aggregation

The term "prion", originally defined by Prusiner, refers to a 'proteinaceous infectious particle' that causes degeneration of the CNS [110]. In addition to prion, other misfolded proteins, most notably Aβ, tau, α-synuclein and TAR DNA-binding protein 43 (TDP43), also act as proteopathic seeds to template the physiological alterations of the same protein, transmit between cells and spread to neuroanatomical connected regions, described as the 'prion-like property' [110].

The natively unfolded and highly soluble tau monomer contains a minimal content of ordered secondary structure, which shows little tendency to undergo aggregation [111]. An initial disruption likely changes tau conformation to obtain more β-sheet structures required for the interaction between monomers [94]. Two hexapeptide motifs of tau, ^{275}VQIINK^{280} and ^{306}VQIVYK^{311}, are crucial for the conformational switch and filament assembly [112]. As the intrinsic cause of fibrillation, the aggregation-prone sequence elements are found in multiple tauopathies (**Table 3**) [10]. Once the partially folded tau monomers are stimulated, they may sequentially aggregate to form dimmers, soluble oligomers, and eventually the insoluble PHFs and NTFs [113]. The microtubule-binding domains that contain β-sheet structures assemble into the rigid core of PHFs, while the N- and C-terminus of tau protein form a 'fuzzy coat' surrounding the core [99]. All six isoforms of CNS tau are detected in AD PHFs [114]. Oligomeric tau and PHF-tau isolated from AD brains can serve as prion-like seeds to induce the aggregation of normal tau both in vitro and in vivo, while monomeric heat-stable and straight filament (SF)-tau showed limited prion-like properties. However, AD O-tau is the most potent toxic species to induce pathological tau aggregation and propagation in vivo [80, 96, 115–117].

Tau fibrils isolated from the brains of patients with different tauopathies show disease-specific folding. In some cases, even in the same disease, the protofilaments of tau pack in distinct ways to form different polymorphs although the conformation of tau monomer is relatively preserved [10, 118]. Tau strains from different sources exhibit variant ability in inducing aggregation in vivo. For example, both the structures of PHF and SF in AD contain eight β-sheets in a C-shaped fold, but the intermolecular organization between the two kinds of protofilament is different [119]. PHF-tau, but not SF-tau, dramatically seeds tau aggregation in vitro and triggers the propagation of tau pathology in vivo [116]. Furthermore, the induced pathology in specific cell types is related to the origin of tau species. Injection of AD pathological tau to mouse brain induces pathology only in neurons, while injection of CBD or PSP pathological tau gives rise to pathology in neurons as well as astrocytes and oligodendrocytes [120].

Truncation and hyperphosphorylation are two major PMTs that contribute to tau aggregation. Truncation of tau may expose the microtubule-binding repeats that are responsible for aggregation. Tau truncation alone is sufficient to induce site-specific phosphorylation and self-aggregates [97]. Among various truncations of tau, deletion of the first 150 aa and the last 50 aa (tau151–391) promotes its pathological characteristics most significantly. Compared with full-length tau, tau151–391 is more prone to phosphorylation, self-aggregation and seeded aggregation by AD O-tau [96]. Hyperphosphorylation alone is not sufficient to induce tau aggregation, but only when it occurs together with truncation [121]. Phosphorylation at several sites (e.g. pT231, pS235 and pS262) flanking the microtubule-binding domains inhibits tau-mediated microtubule assembly and facilitates tau aggregation into PHFs [122]. Interestingly, the phosphorylation levels of these sites are positively correlated to the seeding capacity of tau species isolated from AD patients. Notably, the effect of phosphorylation on tau seeding activity seems to be site-specific. For instance, phosphorylation levels at pS198/S199/S202 and pS400/T403/S404 show a negative correlation with tau seeding activity [72].

Mutation, dysregulation of tau alternative splicing and local contextual factors are also risks for tau aggregation [22]. Moreover, some chemical factors with strong negative charges (e.g. heparin, RNA, dextran sulphate and arachidonic acid) can induce tau aggregation in vitro [10]. Nevertheless, it should be pointed out that none of the fibrils synthesized in vitro resemble patient-derived fibrils so far, because the structures of chemical-induced tau filaments are quite different from those in diseases [118]. In addition to neurons, microglias also take up both soluble and insoluble tau [123] and are involved in promoting the spread of tau pathology [124].

5.2 Propagation of tau pathology

Once formed in the coeruleus/subcoeruleus region, the prion-like tau seeds are transmitted from a "donor cell" to a "recipient cell" to template more aggregation, and progressively spread the pathology along synaptically connected neurons to large-scale human brain networks. Propagation of tau pathology involves the following steps: 1) uptake of tau seeds, 2) seeded aggregation, 3) secretion of new seeds and 4) transcellular transmission of the toxic seeds [125].

The proteopathic seeds released from neurons can be engulfed by interconnected neurons at the somatodendritic compartment and the axon terminals, and are transported both anterogradely and retrogradely [126]. Endocytosis is the primary pathway for the internalization of proteopathic seeds, although the specific mechanisms vary [120]. It has been reported that pathological tau could be taken up by macropinocytosis, heparan sulfate proteoglycans (HSPGs)-dependent endocytosis, clathrin-mediated endocytosis, phagocytosis and tunneling nanotubes (TNT)-dependent direct intercellular transport, of which HSPGs-mediated endocytosis is the most predominant way [127]. The mechanisms of tau uptake are mainly similar for both vesicle-bound and free proteins [126]. Diverse tau strains display different uptake efficiency. Tau trimers are the minimal fragment that can be spontaneously internalized by primary neurons, while no clear upper limit is observed [128]. However, the soluble HMW-tau isolated from AD brain is the most efficiently internalized species [129].

After internalization, tau seeds are transported to the endo-lysosomal system, and some of them are degraded in the lysosomes [130]. Due to age-related dysfunction or other unknown mechanisms [131], a part of the seeds disrupt the endosomal vesicle and enter the cytoplasm, where they template amplification of new fibrils [132]. The danger receptor galectin-8 could protect against the release of seeds by monitoring endomembrane integrity and activating autophagy [133].

Tau contains no signal peptides. But it is secreted into the culture medium, interstitial fluid or CSF in a monomeric and/or truncated non-phosphorylated form lacking the C-terminal portion in physiological conditions [125], implying a potential physiologic role. Nonetheless, the transmission of tau monomer is unlikely to mediate transcellular propagation of tau pathology [134].

The proteopathic tau seeds are released from neurons through multiple pathways, including exocytosis [135], exosome [136], synaptic vesicles [137], presynapse membrane penetration [138] or even direct translocation across neurons [139]. Truncation and hyperphosphorylation of tau or increased synaptic activity facilitate its secretion [94]. The released proteopathic seeds are taken up by postsynaptic neurons and subsequently propagate tau pathology in the interconnected cells.

6. Tau-based therapeutics strategies

So far, the US Food and Drug Administration (FDA) has granted seven prescription drugs for the treatment of AD. Three of these drugs, Donepezil (Eisai Co., Ltd., Pfizer), Galantamine (Janssen, Ortho-McNeil Pharmaceutical, Sanochemia Pharmazeutika, Shire, Takeda Pharmaceutical Company) and Rivastigmine (Novartis Pharmaceuticals Corporation), are cholinesterase inhibitors that prevent the breakdown of acetylcholine in the brain. Rivastigmine has also been approved for the treatment of PD. Applications of Donepezil in the treatment of dementia with Lewy bodies, Down's syndrome and PD are under clinical trial. The forth drug Memantine (Forest Laboratories, Inc., H. Lundbeck, Merz Pharma) is a non-competitive N-methyl-D-aspartate (NMDA) receptor antagonist. The fifth drug Suvorexant (Merck), an orexin receptor antagonist, is approved for treating sleep disorders in AD. The sixth drug Tacrine is a reversible acetylcholinesterase inhibitor, but has been discontinued because of its hepatotoxicity (Pfizer, Shionogi Pharma). In 2021, Aduhelm (Aducanumab), a human IgG1 monoclonal antibody against Aβ (Biogen, Neurimmune), was approved as the first immunotherapy for AD. However, most therapies exhibit limited benefits and do not prevent or slow down the progression of the disease [122].

Based on the extensive understanding of AD pathogenesis, a large number of therapies targeting tau have been developed recently. To data, among the 137 active therapeutic clinical trials for AD, FTD and PSP, 17 targeted tau (https://www.alzforum.org/therapeutics).

6.1 Immunotherapy

Although the mechanism is not fully elucidated, tau immunotherapy, either active or passive, shows protective effects on tau pathology and cognitive performance in AD model animals and is consequently becoming an essential strategy in the development of AD therapies [17]. Two vaccines for active immunotherapy of AD, AADvac-1 which is a synthetic peptide of tau (aa 294–305) and ACI-35 which is a phosphorylated tau peptide containing pS396/S404 are now in clinical trials (**Table 4**) [140]. Since active immunization may increase the risk of autoimmune reaction and other disadvantages [141], passive immunization currently accounts for the majority of the immunotherapy for tau (**Table 4**).

Passive immunotherapy is a short-term immunization administered by continuous injection of antibody that is designed specifically to pathological epitopes. Intravenously injection of tau antibody decreases both Aβ and tau pathologies in animal models, implying a therapeutic potential in the treatment of AD and related tauopathies [142]. There are 8 specific tau antibodies in clinical trials up to now. Their targets mainly focus on the N-terminus (ABBV-8E12 and Semorinemab), phosphorylation sites (Lu AF87908 and JNJ-63733657) or microtubule-binding region (Bepranemab and E2814) of tau protein (**Table 4**). Some antibodies (e.g. Bepranemab) also show potential in the treatment of other neurodegenerative diseases

Approach	Agent	Mechanism of action	Sponsor	Clinical trial	Status
Active	AADvac1	A synthetic peptide derived from tau	Axon Neuroscience SE	Phase 2	Completed
Active	ACI-35.030 JACI-35.054	Synthetic tau fragment phosphorylated at pS396/S404 anchored into a lipid bilayer	AC Immune SA	Phase 1 Phase 2	Recruiting
Passive	ABBV-8E12	A monoclonal antibody recognizes tau's N-terminus	AbbVie	Phase 2	Completed
Passive	BIIB076	A monoclonal anti-tau antibody	Biogen	Phase 1	Completed
Passive	Semorinemab	A monoclonal antibody recognizes tau's N-terminus	Genentech, Inc.	Phase 2	Completed
Passive	LY3303560	A humanized antibody against soluble tau aggregates	Eli Lilly and Company	Phase 2	Active, not recruiting
Passive	Lu AF87908	A monoclonal antibody recognizes phosphorylated tau	H. Lundbeck A/S	Phase 1	Recruiting
Passive	JNJ-63733657	A monoclonal antibody recognizes phosphorylated tau	Janssen Research & Development, LLC	Phase 2	Recruiting
Passive	Bepranemab	A monoclonal antibody recognizes tau235–250	UCB Biopharma SRL	Phase 2	Recruiting
Passive	E2814	A monoclonal antibody recognizes an HVPGG epitope of tau	Eisai Inc.	Phase 1 Phase 2	Recruiting
PTM	Salsalate	Inhibit tau acetylation	Adam Boxer	Phase 1	Active, not recruiting
PTM	LY3372689	A inhibitor of the O-GlcNAcase	Eli Lilly and Company	Phase 1	Completed
PTM	ASN51	A small-molecule inhibitor of O-GlycNAcase	Asceneuron Pty Ltd.	Phase 1	Recruiting
ASO	BIIB080	An antisense oligonucleotide (ASO) targeting tau expression	Ionis Pharmaceuticals, Inc.	Phase 1 Phase 2	Active, not recruiting

Data from: https://clinicaltrials.gov/, https://www.alzforum.org/therapeutics/

Table 4.
Tau-based therapies for AD.

like PSP (https://www.alzforum.org/therapeutics) [143]. Of course, tau-based passive immunization also faces some challenges, of which the key problem is how to deliver the antibodies more efficiently through the blood–brain-barrier (BBB) into specific brain region with tau pathology [144].

6.2 Therapies targeting tau PTMs

Acetylation at specific sites of tau is shown to inhibit the degradation of hyperphosphorylated tau, obstruct synaptic plasticity and promote cognitive impairment in AD mouse models [91]. Salsalate, a non-acetylated dimer of salicylic acid commonly used as a non-steroidal anti-inflammatory drug, can inhibit acetyltransferase p300-induced tau acetylation, thus enhancing tau turnover and reducing tau levels [145]. O-GlcNAcylation negatively regulates tau phosphorylation by competing for the phosphorylation sites [146]. Two small molecular inhibitors of O-GlycNAcase (LY3372689 and ASN51) are currently undergoing clinical trials.

Hyperphosphorylation is the crucial PTM that determines the propagation of tau pathology. Inhibition of tau hyperphosphorylation has long been considered as a potential therapeutic strategy. Tau phosphorylation can be modulated by the balance between protein kinases and phosphatases. Nasal insulin and the GSK-3 inhibitor Lithium that inhibit tau phosphorylation via activating PI3K signaling, and PP2A activator Metformin aimed at tau dephosphorylation are currently under development or evaluation in clinical trials [122].

6.3 Other therapies

Knockout of MAPT gene induces no obvious phenotype except for behavioral deficits in aged mice [147]. Reducing the levels of endogenous tau shows protection against cognitive impairments and behavioral abnormalities in AD mice [148]. BIIB080, the first antisense oligonucleotide (ASO) targeting the translation of tau mRNAs, has just started Phase 1 and Phase 2 clinical trials. Recently, a selective protein degradation approach achieved by the proteolysis targeting chimeras (PROTACs) is utilized to decrease tau protein in the brain [82, 149]. PROTACs form a ternary complex with the target protein and ubiquitin E3 ligase. E3 ubiquitin ligase stimulates polyubiquitination of the targets and facilitates its following recognition and degradation by the 26S proteasome [149]. PROTACs targeting tau remarkably decreased tau levels and improved synaptic and cognitive functions in wildtype and AD mice [150].

In the past few years, inhibitors of tau aggregation were considered as potential therapies for AD for their roles in preventing the prion-like seeding and propagation of tau pathology. Unfortunately, the clinical trials of corresponding small molecular drugs, such as methylene blue, Rember TM, and LMTX, did not show the expected effect and are thus discontinued. NPT088, a fusion protein consisting of human-IgG 1-Fc and an active fragment binds to and remodels misfolded aggregates of tau, also exhibited no effect on brain plaques, tau aggregates or AD symptoms [151].

What's more, the disruption of microtubules is one of the main consequences of tau-induced neurotoxicity. Therefore, stabilization of microtubules may also be a potential therapeutic approach associated with tau. However, the efficacy of microtubule stabilizers (e.g. TPI-287) in the treatment of AD still requires further evaluation.

7. Conclusions and perspectives

In summary, tau-mediated microtubule dynamics and assembly play essential roles in neuronal transport and the maintenance of synaptic structure and function. In neurodegenerative diseases, tau undergoes a series of pathological changes, including mutation, abnormal alternative splicing, abnormal PTMs and prion-like seeding and propagation. Certain benefits of therapeutic approaches targeting pathological tau have emerged in the treatment of AD and related tauopathies. Recently, proteomics results indicate significant heterogeneity of tau pathology in different patients, raising the possibility that personalized approaches according to the biochemical characteristics of tau may achieve better therapeutic effects [10]. It is also worth noting that tau interacts with other risk factors of neurodegenerative diseases, such as Aβ [152], apolipoprotein E [153], α-synuclein [154], metal dysregulation [155], defective mitophagy [156], stress and inflammation [157] and so on. Therefore, further investigation of the comprehensive map of tau interactions will better reveal the pathogenesis of neurodegenerative diseases.

Acknowledgements

This work was supported by Nantong University, New York State Office for People Developmental Disabilities, the Neural Regeneration Co-innovation Center of Jiangsu Province and grant from the National Natural Science Foundation of China (Grants 81872853).

Conflict of interest

The authors declare no conflict of interest.

Author details

Dandan Chu[1*] and Fei Liu[2*]

1 Key Laboratory of Neuroregeneration of Jiangsu and Ministry of Education, NMPA Key Laboratory for Research and Evaluation of Tissue Engineering Technology Products, Co-innovation Center of Neuroregeneration, Nantong University, Nantong, Jiangsu, China

2 Department of Neurochemistry, Inge Grundke-Iqbal Research Floor, New York State Institute for Basic Research in Developmental Disabilities, New York, USA

*Address all correspondence to: chudd@ntu.edu.cn and feiliu63@hotmail.com

References

[1] Fjell AM, Mcevoy L, Holland D, Dale AM, Walhovd KB. What is normal in normal aging? Effects of aging, amyloid and Alzheimer's disease on the cerebral cortex and the hippocampus. Progress in Neurobiology. 2014;**117**:20-40

[2] Veitch DP, Weiner MW, Aisen PS, Beckett LA, Cairns NJ, Green RC, et al. Understanding disease progression and improving Alzheimer's disease clinical trials: Recent highlights from the Alzheimer's Disease Neuroimaging Initiative. Alzheimers Dement. 2019;**15**: 106-152

[3] Berger T, Lee H, Young AH, Aarsland D, Thuret S. Adult hippocampal neurogenesis in major depressive disorder and Alzheimer's disease. Trends in Molecular Medicine. 2020;**26**:803-818

[4] Weingarten MD, Lockwood AH, Hwo SY, Kirschner MW. A protein factor essential for microtubule assembly. Proceedings of the National Academy of Sciences of the United States of America. 1975;**72**:1858-1862

[5] Hu W, Wu F, Zhang Y, Gong CX, Iqbal K, Liu F. Expression of Tau pathology-related proteins in different brain regions: A molecular basis of tau pathogenesis. Frontiers in Aging Neuroscience. 2017;**9**:311

[6] Iqbal K, Wisniewski HM, Shelanski ML, Brostoff S, Liwnicz BH, Terry RD. Protein changes in senile dementia. Brain Research. 1974;**77**: 337-343

[7] Grundke-Iqbal I, Iqbal K, Quinlan M, Tung YC, Zaidi MS, Wisniewski HM. Microtubule-associated protein tau. A component of Alzheimer paired helical filaments. The Journal of Biological Chemistry. 1986;**261**:6084-6089

[8] Gotz J, Halliday G, Nisbet RM. Molecular Pathogenesis of the Tauopathies. Annual Review of Pathology. 2019;**14**:239-261

[9] Forrest SL, Kril JJ, Halliday GM. Cellular and regional vulnerability in frontotemporal tauopathies. Acta Neuropathologica. 2019;**138**:705-727

[10] Vaquer-Alicea J, Diamond MI, Joachimiak LA. Tau strains shape disease. Acta Neuropathologica. 2021;**142**:57-71

[11] WHO. 2017. http://www.who.int/en/news-room/fact-sheets/detail/dementia

[12] Kent SA, Spires-Jones TL, Durrant CS. The physiological roles of tau and Abeta: implications for Alzheimer's disease pathology and therapeutics. Acta Neuropathologica. 2020;**140**:417-447

[13] Braak H, Braak E. Neuropathological stageing of Alzheimer-related changes. Acta Neuropathologica. 1991;**82**:239-259

[14] Braak H, Braak E. Staging of Alzheimer's disease-related neurofibrillary changes. Neurobiology of Aging. 1995;**16**:271-278; discussion 278-284

[15] Long JM, Holtzman DM. Alzheimer disease: An update on pathobiology and treatment strategies. Cell. 2019;**179**: 312-339

[16] Ittner A, Ittner LM. Dendritic Tau in Alzheimer's Disease. Neuron. 2018;**99**: 13-27

[17] Chang CW, Shao E, Mucke L. Tau: Enabler of diverse brain disorders and target of rapidly evolving therapeutic strategies. Science. 2021;**371**:1-10

[18] Goedert M, Spillantini MG, Jakes R, Rutherford D, Crowther RA. Multiple isoforms of human microtubule-associated protein tau: sequences and

localization in neurofibrillary tangles of Alzheimer's disease. Neuron. 1989;**3**: 519-526

[19] Fischer I, Baas PW. Resurrecting the mysteries of Big Tau. Trends in Neurosciences. 2020;**43**:493-504

[20] Kosik KS, Orecchio LD, Bakalis S, Neve RL. Developmentally regulated expression of specific tau sequences. Neuron. 1989;**2**:1389-1397

[21] Qian W, Liu F. Regulation of alternative splicing of tau exon 10. Neuroscience Bulletin. 2014;**30**:367-377

[22] Chu D, Liu F. Pathological changes of Tau related to Alzheimer's disease. ACS Chemical Neuroscience. 2019;**10**:931-944

[23] Rosler TW, Tayaranian Marvian A, Brendel M, Nykanen NP, Hollerhage M, Schwarz SC, et al. Four-repeat tauopathies. Progress in Neurobiology. 2019;**180**:101644

[24] Chambers CB, Lee JM, Troncoso JC, Reich S, Muma NA. Overexpression of four-repeat tau mRNA isoforms in progressive supranuclear palsy but not in Alzheimer's disease. Annals of Neurology. 1999;**46**:325-332

[25] Ginsberg SD, Che S, Counts SE, Mufson EJ. Shift in the ratio of three-repeat tau and four-repeat tau mRNAs in individual cholinergic basal forebrain neurons in mild cognitive impairment and Alzheimer's disease. Journal of Neurochemistry. 2006;**96**:1401-1408

[26] Togo T, Akiyama H, Iseki E, Uchikado H, Kondo H, Ikeda K, et al. Immunohistochemical study of tau accumulation in early stages of Alzheimer-type neurofibrillary lesions. Acta Neuropathologica. 2004;**107**: 504-508

[27] Espinoza M, de Silva R, Dickson DW, Davies P. Differential incorporation of tau isoforms in Alzheimer's disease. Journal of Alzheimer's Disease. 2008;**14**:1-16

[28] Wang Y, Mandelkow E. Tau in physiology and pathology. Nature Reviews. Neuroscience. 2016;**17**:5-21

[29] Caillet-Boudin ML, Buee L, Sergeant N, Lefebvre B. Regulation of human MAPT gene expression. Molecular Neurodegeneration. 2015;**10**:28

[30] Sun W, Jia J. The +347 C promoter allele up-regulates MAPT expression and is associated with Alzheimer's disease among the Chinese Han. Neuroscience Letters. 2009;**450**:340-343

[31] Baker M, Litvan I, Houlden H, Adamson J, Dickson D, Perez-Tur J, et al. Association of an extended haplotype in the tau gene with progressive supranuclear palsy. Human Molecular Genetics. 1999;**8**:711-715

[32] Latimer CS, Lucot KL, Keene CD, Cholerton B, Montine TJ. Genetic insights into Alzheimer's disease. Annual Review of Pathology. 2021;**16**:351-376

[33] Mandelkow EM, Biernat J, Drewes G, Gustke N, Trinczek B, Mandelkow E. Tau domains, phosphorylation, and interactions with microtubules. Neurobiology of Aging. 1995;**16**:355-362; discussion 362-353

[34] Kahlson MA, Colodner KJ. Glial Tau pathology in tauopathies: Functional consequences. Journal of Experimental Neuroscience. 2015;**9**:43-50

[35] Wu R, Gu J, Zhou D, Tung YC, Jin N, Chu D, et al. Seeding-competent Tau in gray matter versus white matter of Alzheimer's disease brain. Journal of Alzheimer's Disease. 2021;**79**:1647-1659

[36] Bancher C, Lassmann H, Budka H, Grundke-Iqbal I, Iqbal K, Wiche G, et al. Neurofibrillary tangles in Alzheimer's disease and progressive supranuclear palsy: Antigenic similarities and

differences. Microtubule-associated protein tau antigenicity is prominent in all types of tangles. Acta Neuropathologica. 1987;**74**:39-46

[37] Ittner LM, Ke YD, Delerue F, Bi M, Gladbach A, van Eersel J, et al. Dendritic function of tau mediates amyloid-beta toxicity in Alzheimer's disease mouse models. Cell. 2010;**142**:387-397

[38] Zempel H, Mandelkow E. Lost after translation: Missorting of Tau protein and consequences for Alzheimer disease. Trends in Neurosciences. 2014;**37**:721-732

[39] Zempel H, Dennissen FJA, Kumar Y, Luedtke J, Biernat J, Mandelkow EM, et al. Axodendritic sorting and pathological missorting of Tau are isoform-specific and determined by axon initial segment architecture. The Journal of Biological Chemistry. 2017;**292**:12192-12207

[40] Li C, Gotz J. Somatodendritic accumulation of Tau in Alzheimer's disease is promoted by Fyn-mediated local protein translation. The EMBO Journal. 2017;**36**:3120-3138

[41] Kobayashi S, Tanaka T, Soeda Y, Almeida OFX, Takashima A. Local somatodendritic translation and hyperphosphorylation of Tau protein triggered by AMPA and NMDA receptor stimulation. eBioMedicine. 2017;**20**: 120-126

[42] Sultan A, Nesslany F, Violet M, Begard S, Loyens A, Talahari S, et al. Nuclear tau, a key player in neuronal DNA protection. The Journal of Biological Chemistry. 2011;**286**:4566-4575

[43] Baas PW, Rao AN, Matamoros AJ, Leo L. Stability properties of neuronal microtubules. Cytoskeleton (Hoboken). 2016;**73**:442-460

[44] Guzman-Martinez L, Tapia JP, Farias GA, Gonzalez A, Estrella M, Maccioni RB. The Alz-tau biomarker for Alzheimer's disease: Study in a caucasian population. Journal of Alzheimer's Disease. 2019;**67**:1181-1186

[45] Kellogg EH, Hejab NMA, Poepsel S, Downing KH, DiMaio F, Nogales E. Near-atomic model of microtubule-tau interactions. Science. 2018;**360**:1242-1246

[46] Montejo de Garcini E, de la Luna S, Dominguez JE, Avila J. Overexpression of tau protein in COS-1 cells results in the stabilization of centrosome-independent microtubules and extension of cytoplasmic processes. Molecular and Cellular Biochemistry. 1994;**130**:187-196

[47] Kadavath H, Hofele RV, Biernat J, Kumar S, Tepper K, Urlaub H, et al. Tau stabilizes microtubules by binding at the interface between tubulin heterodimers. Proceedings of the National Academy of Sciences of the United States of America. 2015;**112**:7501-7506

[48] Yin X, Jin N, Shi J, Zhang Y, Wu Y, Gong CX, et al. Dyrk1A overexpression leads to increase of 3R-tau expression and cognitive deficits in Ts65Dn down syndrome mice. Scientific Reports. 2017;**7**:619

[49] Goedert M, Jakes R. Expression of separate isoforms of human tau protein: correlation with the tau pattern in brain and effects on tubulin polymerization. The EMBO Journal. 1990;**9**:4225-4230

[50] Hernandez-Vega A, Braun M, Scharrel L, Jahnel M, Wegmann S, Hyman BT, et al. Local nucleation of microtubule bundles through Tubulin concentration into a condensed Tau Phase. Cell Reports. 2017;**20**:2304-2312

[51] Qiang L, Sun X, Austin T, Muralidharan H, Jean L, Liu M, et al. Tau does not stabilize axonal microtubules but rather enables them to have long labile domains. Current Biology. 2018;**28**:1-9

[52] Kanaan NM, Morfini GA, LaPointe NE, Pigino GF, Patterson KR,

Song Y, et al. Pathogenic forms of tau inhibit kinesin-dependent axonal transport through a mechanism involving activation of axonal phosphotransferases. The Journal of Neuroscience. 2011;**31**:9858-9868

[53] Tapia-Rojas C, Cabezas-Opazo F, Deaton CA, Vergara EH, Johnson GVW, Quintanilla RA. It's all about tau. Progress in Neurobiology. 2019;**175**:54-76

[54] Chen Y, Fu AKY, Ip NY. Synaptic dysfunction in Alzheimer's disease: Mechanisms and therapeutic strategies. Pharmacology & Therapeutics. 2019; **195**:186-198

[55] TO A, Qiang L, Pw B. Mechanisms of neuronal microtubule loss in Alzheimer's disease. In: Neuroprotection in Alzheimer's Disease. Amsterdam: Elsevier Inc; 2017

[56] Niewidok B, Igaev M, Sundermann F, Janning D, Bakota L, Brandt R. Presence of a carboxy-terminal pseudorepeat and disease-like pseudohyperphosphorylation critically influence tau's interaction with microtubules in axon-like processes. Molecular Biology of the Cell. 2016;**27**: 3537-3549

[57] Stern JL, Lessard DV, Hoeprich GJ, Morfini GA, Berger CL. Phospho-regulation of Tau modulates inhibition of kinesin-1 motility. Molecular Biology of the Cell. 2017;**28**:1079-1087

[58] Swanson E, Breckenridge L, McMahon L, Som S, McConnell I, Bloom GS. Extracellular Tau oligomers induce invasion of endogenous Tau into the somatodendritic compartment and axonal transport dysfunction. Journal of Alzheimer's Disease. 2017;**58**:803-820

[59] Moreno H, Morfini G, Buitrago L, Ujlaki G, Choi S, Yu E, et al. Tau pathology-mediated presynaptic dysfunction. Neuroscience. 2016;**325**: 30-38

[60] Higuchi M, Zhang B, Forman MS, Yoshiyama Y, Trojanowski JQ, Lee VM. Axonal degeneration induced by targeted expression of mutant human tau in oligodendrocytes of transgenic mice that model glial tauopathies. The Journal of Neuroscience. 2005;**25**:9434-9443

[61] Muller R, Heinrich M, Heck S, Blohm D, Richter-Landsberg C. Expression of microtubule-associated proteins MAP2 and tau in cultured rat brain oligodendrocytes. Cell and Tissue Research. 1997;**288**:239-249

[62] Zlokovic BV. The blood-brain barrier in health and chronic neurodegenerative disorders. Neuron. 2008;**57**:178-201

[63] Forman MS, Lal D, Zhang B, Dabir DV, Swanson E, Lee VM, et al. Transgenic mouse model of tau pathology in astrocytes leading to nervous system degeneration. The Journal of Neuroscience. 2005;**25**:3539-3550

[64] Wang P, Ye Y. Filamentous recombinant human Tau activates primary astrocytes via an integrin receptor complex. Nature Communications. 2021;**12**:95

[65] Asai H, Ikezu S, Tsunoda S, Medalla M, Luebke J, Haydar T, et al. Depletion of microglia and inhibition of exosome synthesis halt tau propagation. Nature Neuroscience. 2015;**18**:1584-1593

[66] Yin H, Kuret J. C-terminal truncation modulates both nucleation and extension phases of tau fibrillization. FEBS Letters. 2006;**580**:211-215

[67] Barthelemy NR, Fenaille F, Hirtz C, Sergeant N, Schraen-Maschke S, Vialaret J, et al. Tau protein quantification in human cerebrospinal fluid by targeted mass spectrometry at high sequence coverage provides insights into its primary structure heterogeneity. Journal of Proteome Research. 2016;**15**:667-676

[68] Sato C, Barthelemy NR, Mawuenyega KG, Patterson BW, Gordon BA, Jockel-Balsarotti J, et al. Tau kinetics in neurons and the human central nervous system. Neuron. 2018;**98**:861-864

[69] Hitt BD, Vaquer-Alicea J, Manon VA, Beaver JD, Kashmer OM, Garcia JN, et al. Ultrasensitive tau biosensor cells detect no seeding in Alzheimer's disease CSF. Acta Neuropathologica Communications. 2021;**9**:99

[70] Wesseling H, Mair W, Kumar M, Schlaffner CN, Tang S, Beerepoot P, et al. Tau PTM profiles identify patient heterogeneity and stages of Alzheimer's disease. Cell. 2020;**183**(1699-1713):e1613

[71] Iqbal K, Liu F, Gong CX. Tau and neurodegenerative disease: the story so far. Nature Reviews. Neurology. 2016;**12**: 15-27

[72] Dujardin S, Commins C, Lathuiliere A, Beerepoot P, Fernandes AR, Kamath TV, et al. Tau molecular diversity contributes to clinical heterogeneity in Alzheimer's disease. Nature Medicine. 2020;**26**:1256-1263

[73] Morris M, Knudsen GM, Maeda S, Trinidad JC, Ioanoviciu A, Burlingame AL, et al. Tau post-translational modifications in wild-type and human amyloid precursor protein transgenic mice. Nature Neuroscience. 2015;**18**:1183-1189

[74] Li L, Jiang Y, Hu W, Tung YC, Dai C, Chu D, et al. Pathological alterations of Tau in Alzheimer's disease and 3xTg-AD mouse brains. Molecular Neurobiology. 2019;**56**:6168-6183

[75] Kopke E, Tung YC, Shaikh S, Alonso AC, Iqbal K, Grundke-Iqbal I. Microtubule-associated protein tau. Abnormal phosphorylation of a non-paired helical filament pool in Alzheimer disease. The Journal of Biological Chemistry. 1993;**268**:24374-24384

[76] Mandelkow E, von Bergen M, Biernat J, Mandelkow EM. Structural principles of tau and the paired helical filaments of Alzheimer's disease. Brain Pathology. 2007;**17**:83-90

[77] Iqbal K, Alonso Adel C, Grundke-Iqbal I. Cytosolic abnormally hyperphosphorylated tau but not paired helical filaments sequester normal MAPs and inhibit microtubule assembly. Journal of Alzheimer's Disease. 2008;**14**: 365-370

[78] Drummond E, Pires G, MacMurray C, Askenazi M, Nayak S, Bourdon M, et al. Phosphorylated tau interactome in the human Alzheimer's disease brain. Brain. 2020;**143**:2803-2817

[79] Liu F, Grundke-Iqbal I, Iqbal K, Gong CX. Contributions of protein phosphatases PP1, PP2A, PP2B and PP5 to the regulation of tau phosphorylation. The European Journal of Neuroscience. 2005;**22**:1942-1950

[80] Hu W, Zhang X, Tung YC, Xie S, Liu F, Iqbal K. Hyperphosphorylation determines both the spread and the morphology of tau pathology. Alzheimers Dement. 2016;**12**:1066-1077

[81] Wang Y, Zhang Y, Hu W, Xie S, Gong CX, Iqbal K, et al. Rapid alteration of protein phosphorylation during postmortem: implication in the study of protein phosphorylation. Scientific Reports. 2015;**5**:15709

[82] Schmidt MF, Gan ZY, Komander D, Dewson G. Ubiquitin signalling in neurodegeneration: mechanisms and therapeutic opportunities. Cell Death and Differentiation. 2021;**28**:570-590

[83] Abreha MH, Dammer EB, Ping L, Zhang T, Duong DM, Gearing M, et al. Quantitative analysis of the brain ubiquitylome in Alzheimer's Disease. Proteomics. 2018;**18**:e1800108

[84] Kontaxi C, Piccardo P, Gill AC. Lysine-directed post-translational

modifications of Tau protein in Alzheimer's disease and related tauopathies. Frontiers in Molecular Biosciences. 2017;**4**:56

[85] Dickey CA, Kamal A, Lundgren K, Klosak N, Bailey RM, Dunmore J, et al. The high-affinity HSP90-CHIP complex recognizes and selectively degrades phosphorylated tau client proteins. The Journal of Clinical Investigation. 2007;**117**:648-658

[86] Tan JM, Wong ES, Kirkpatrick DS, Pletnikova O, Ko HS, Tay SP, et al. Lysine 63-linked ubiquitination promotes the formation and autophagic clearance of protein inclusions associated with neurodegenerative diseases. Human Molecular Genetics. 2008;**17**:431-439

[87] Grune T, Botzen D, Engels M, Voss P, Kaiser B, Jung T, et al. Tau protein degradation is catalyzed by the ATP/ubiquitin-independent 20S proteasome under normal cell conditions. Archives of Biochemistry and Biophysics. 2010;**500**:181-188

[88] Galves M, Rathi R, Prag G, Ashkenazi A. Ubiquitin signaling and degradation of aggregate-prone proteins. Trends in Biochemical Sciences. 2019;**44**: 872-884

[89] Thibaudeau TA, Anderson RT, Smith DM. A common mechanism of proteasome impairment by neuro-degenerative disease-associated oligomers. Nature Communications. 2018;**9**:1097

[90] Cohen TJ, Constance BH, Hwang AW, James M, Yuan CX. Intrinsic Tau acetylation is coupled to auto-proteolytic Tau fragmentation. PLoS One. 2016;**11**:e0158470

[91] Tracy TE, Sohn PD, Minami SS, Wang C, Min SW, Li Y, et al. Acetylated Tau obstructs KIBRA-mediated signaling in synaptic plasticity and promotes tauopathy-related memory loss. Neuron. 2016;**90**:245-260

[92] Cook C, Carlomagno Y, Gendron TF, Dunmore J, Scheffel K, Stetler C, et al. Acetylation of the KXGS motifs in tau is a critical determinant in modulation of tau aggregation and clearance. Human Molecular Genetics. 2014;**23**:104-116

[93] Zhou Y, Shi J, Chu D, Hu W, Guan Z, Gong CX, et al. Relevance of phosphorylation and truncation of Tau to the etiopathogenesis of Alzheimer's disease. Frontiers in Aging Neuroscience. 2018;**10**:27

[94] Quinn JP, Corbett NJ, Kellett KAB, Hooper NM. Tau proteolysis in the pathogenesis of tauopathies: Neurotoxic fragments and novel biomarkers. Journal of Alzheimer's Disease. 2018;**63**:13-33

[95] Zhang Z, Song M, Liu X, Kang SS, Kwon IS, Duong DM, et al. Cleavage of tau by asparagine endopeptidase mediates the neurofibrillary pathology in Alzheimer's disease. Nature Medicine. 2014;**20**:1254-1262

[96] Gu J, Xu W, Jin N, Li L, Zhou Y, Chu D, et al. Truncation of tau selectively facilitates its pathological activities. The Journal of Biological Chemistry. 2020;**295**:13812-13828

[97] Kovacech B, Novak M. Tau truncation is a productive posttranslational modification of neurofibrillary degeneration in Alzheimer's disease. Current Alzheimer Research. 2010;**7**:708-716

[98] Zilkova M, Zilka N, Kovac A, Kovacech B, Skrabana R, Skrabanova M, et al. Hyperphosphorylated truncated protein tau induces caspase-3 independent apoptosis-like pathway in the Alzheimer's disease cellular model. Journal of Alzheimer's Disease. 2011;**23**: 161-169

[99] Fitzpatrick AWP, Falcon B, He S, Murzin AG, Murshudov G, Garringer HJ, et al. Cryo-EM structures of tau

filaments from Alzheimer's disease. Nature. 2017;**547**:185-190

[100] Inekci D, Henriksen K, Linemann T, Karsdal MA, Habib A, Bisgaard C, et al. Serum fragments of Tau for the differential diagnosis of Alzheimer's disease. Current Alzheimer Research. 2015;**12**:829-836

[101] Meredith JE Jr, Sankaranarayanan S, Guss V, Lanzetti AJ, Berisha F, Neely RJ, et al. Characterization of novel CSF Tau and ptau biomarkers for Alzheimer's disease. PLoS One. 2013;**8**:e76523

[102] Amadoro G, Corsetti V, Sancesario GM, Lubrano A, Melchiorri G, Bernardini S, et al. Cerebrospinal fluid levels of a 20-22 kDa NH2 fragment of human tau provide a novel neuronal injury biomarker in Alzheimer's disease and other dementias. Journal of Alzheimer's Disease. 2014;**42**:211-226

[103] Thomas SN, Funk KE, Wan Y, Liao Z, Davies P, Kuret J, et al. Dual modification of Alzheimer's disease PHF-tau protein by lysine methylation and ubiquitylation: A mass spectrometry approach. Acta Neuropathologica. 2012;**123**:105-117

[104] Dorval V, Fraser PE. Small ubiquitin-like modifier (SUMO) modification of natively unfolded proteins tau and alpha-synuclein. The Journal of Biological Chemistry. 2006;**281**:9919-9924

[105] Luo HB, Xia YY, Shu XJ, Liu ZC, Feng Y, Liu XH, et al. SUMOylation at K340 inhibits tau degradation through deregulating its phosphorylation and ubiquitination. Proceedings of the National Academy of Sciences of the United States of America. 2014;**111**:16586-16591

[106] Reyes JF, Geula C, Vana L, Binder LI. Selective tau tyrosine nitration in non-AD tauopathies. Acta Neuropathologica. 2012;**123**:119-132

[107] Reyes JF, Reynolds MR, Horowitz PM, Fu Y, Guillozet-Bongaarts AL, Berry R, et al. A possible link between astrocyte activation and tau nitration in Alzheimer's disease. Neurobiology of Disease. 2008;**31**: 198-208

[108] Braak H, Del Tredici K. The pathological process underlying Alzheimer's disease in individuals under thirty. Acta Neuropathologica. 2011;**121**: 171-181

[109] Boluda S, Iba M, Zhang B, Raible KM, Lee VM, Trojanowski JQ. Differential induction and spread of tau pathology in young PS19 tau transgenic mice following intracerebral injections of pathological tau from Alzheimer's disease or corticobasal degeneration brains. Acta Neuropathologica. 2015;**129**:221-237

[110] Jaunmuktane Z, Brandner S. Invited review: The role of prion-like mechanisms in neurodegenerative diseases. Neuropathology and Applied Neurobiology. 2020;**46**:522-545

[111] Schweers O, Schonbrunn-Hanebeck E, Marx A, Mandelkow E. Structural studies of tau protein and Alzheimer paired helical filaments show no evidence for beta-structure. The Journal of Biological Chemistry. 1994;**269**:24290-24297

[112] Von BM, Barghorn S, Li L, Marx A, Biernat J, Mandelkow EM, et al. Mutations of tau protein in fronto-temporal dementia promote aggregation of paired helical filaments by enhancing local beta-structure. Journal of Biological Chemistry. 2001;**276**: 48165-48174

[113] Mirbaha H, Chen D, Morazova OA, Ruff KM, Sharma AM, Liu X, et al. Inert and seed-competent tau monomers suggest structural origins of aggregation. eLife. 2018;7:1-29

[114] Goedert M, Eisenberg DS, Crowther RA. Propagation of Tau

aggregates and neurodegeneration. Annual Review of Neuroscience. 2017;**40**:189-210

[115] Wu R, Li L, Shi R, Zhou Y, Jin N, Gu J, et al. Dephosphorylation passivates the seeding activity of oligomeric Tau derived from Alzheimer's brain. Frontiers in Molecular Neuroscience. 2021;**14**:631833

[116] Li L, Shi R, Gu J, Tung YC, Zhou Y, Zhou D, et al. Alzheimer's disease brain contains tau fractions with differential prion-like activities. Acta Neuropathologica Communications. 2021;**9**:28

[117] Miao J, Shi R, Li L, Chen F, Zhou Y, Tung YC, et al. Pathological Tau from Alzheimer's brain induces site-specific hyperphosphorylation and SDS- and reducing agent-resistant aggregation of Tau in vivo. Frontiers in Aging Neuroscience. 2019;**11**:34

[118] Li D, Liu C. Hierarchical chemical determination of amyloid polymorphs in neurodegenerative disease. Nature Chemical Biology. 2021;**17**:237-245

[119] Nguyen PH, Ramamoorthy A, Sahoo BR, Zheng J, Faller P, Straub JE, et al. Amyloid oligomers: A joint experimental/computational perspective on Alzheimer's disease, Parkinson's disease, type II diabetes, and amyotrophic lateral sclerosis. Chemical Reviews. 2021;**121**:2545-2647

[120] Peng C, Trojanowski JQ, Lee VM. Protein transmission in neurodegenerative disease. Nature Reviews. Neurology. 2020;**16**:199-212

[121] Cho JH, Johnson GV. Glycogen synthase kinase 3 beta induces caspase-cleaved tau aggregation in situ. The Journal of Biological Chemistry. 2004;**279**:54716-54723

[122] Iqbal K, Liu F, Gong CX. Recent developments with tau-based drug discovery. Expert Opinion on Drug Discovery. 2018;**13**:399-410

[123] Bolos M, Llorens-Martin M, Jurado-Arjona J, Hernandez F, Rabano A, Avila J. Direct evidence of internalization of Tau by Microglia in vitro and in vivo. Journal of Alzheimer's Disease. 2016;**50**:77-87

[124] Maphis N, Xu G, Kokiko-Cochran ON, Jiang S, Cardona A, Ransohoff RM, et al. Reactive microglia drive tau pathology and contribute to the spreading of pathological tau in the brain. Brain. 2015;**138**:1738-1755

[125] Mudher A, Colin M, Dujardin S, Medina M, Dewachter I, Alavi Naini SM, et al. What is the evidence that tau pathology spreads through prion-like propagation? Acta Neuropathologica Communications. 2017;**5**:99

[126] Gibbons GS, Lee VMY, Trojanowski JQ. Mechanisms of cell-to-cell transmission of pathological Tau: A review. JAMA Neurology. 2019;**76**:101-108

[127] Zhao J, Wu H, Tang XQ. Tau internalization: A complex step in tau propagation. Ageing Research Reviews. 2021;**67**:101272

[128] Mirbaha H, Holmes BB, Sanders DW, Bieschke J, Diamond MI. Tau trimers are the minimal propagation unit spontaneously internalized to seed intracellular aggregation. The Journal of Biological Chemistry. 2015;**290**: 14893-14903

[129] Takeda S, Wegmann S, Cho H, DeVos SL, Commins C, Roe AD, et al. Neuronal uptake and propagation of a rare phosphorylated high-molecular-weight tau derived from Alzheimer's disease brain. Nature Communications. 2015;**6**:8490

[130] Wischik CM, Schelter BO, Wischik DJ, Storey JMD, Harrington CR. Modeling prion-like processing of Tau protein in Alzheimer's disease for pharmaceutical

development. Journal of Alzheimer's Disease. 2018;**62**:1287-1303

[131] Ihara Y, Morishima-Kawashima M, Nixon R. The ubiquitin-proteasome system and the autophagic-lysosomal system in Alzheimer disease. In: Cold Spring Harb Perspect Med 2. New York: Cold Spring Harbor Laboratory Press; 2012

[132] Uemura N, Uemura MT, Luk KC, Lee VM, Trojanowski JQ. Cell-to-cell transmission of Tau and alpha-synuclein. Trends in Molecular Medicine. 2020;**26**: 936-952

[133] Falcon B, Noad J, McMahon H, Randow F, Goedert M. Galectin-8-mediated selective autophagy protects against seeded tau aggregation. The Journal of Biological Chemistry. 2018;**293**:2438-2451

[134] Vaquer-Alicea J, Diamond MI. Propagation of protein aggregation in neurodegenerative diseases. Annual Review of Biochemistry. 2019;**88**:785-810

[135] Polanco JC, Li C, Durisic N, Sullivan R, Gotz J. Exosomes taken up by neurons hijack the endosomal pathway to spread to interconnected neurons. Acta Neuropathologica Communications. 2018;**6**:10

[136] Saman S, Kim W, Raya M, Visnick Y, Miro S, Jackson B, et al. Exosome-associated tau is secreted in tauopathy models and is selectively phosphorylated in cerebrospinal fluid in early Alzheimer disease. The Journal of Biological Chemistry. 2012;**287**:3842-3849

[137] DeVos SL, Corjuc BT, Oakley DH, Nobuhara CK, Bannon RN, Chase A, et al. Synaptic Tau seeding precedes Tau pathology in human Alzheimer's disease brain. Frontiers in Neuroscience. 2018; **12**:267

[138] Flach K, Hilbrich I, Schiffmann A, Gartner U, Kruger M, Leonhardt M, et al. Tau oligomers impair artificial membrane integrity and cellular viability. The Journal of Biological Chemistry. 2012;**287**:43223-43233

[139] Katsinelos T, Zeitler M, Dimou E, Karakatsani A, Muller HM, Nachman E, et al. Unconventional secretion mediates the trans-cellular spreading of Tau. Cell Reports. 2018;**23**:2039-2055

[140] Kontsekova E, Zilka N, Kovacech B, Novak P, Novak M. First-in-man tau vaccine targeting structural determinants essential for pathological tau-tau interaction reduces tau oligomerisation and neurofibrillary degeneration in an Alzheimer's disease model. Alzheimer's Research & Therapy. 2014;**6**:44

[141] Winblad B, Graf A, Riviere ME, Andreasen N, Ryan JM. Active immunotherapy options for Alzheimer's disease. Alzheimer's Research & Therapy. 2014;**6**:7

[142] Dai CL, Hu W, Tung YC, Liu F, Gong CX, Iqbal K. Tau passive immunization blocks seeding and spread of Alzheimer hyperphosphorylated Tau-induced pathology in 3 x Tg-AD mice. Alzheimer's Research & Therapy. 2018;**10**:13

[143] Gallardo G, Holtzman DM. Antibody Therapeutics Targeting Abeta and Tau. Cold Spring Harbor Perspectives in Medicine. 2017;7:1-17

[144] Bittar A, Sengupta U, Kayed R. Prospects for strain-specific immunotherapy in Alzheimer's disease and tauopathies. NPJ Vaccines. 2018;**3**:9

[145] Min SW, Chen X, Tracy TE, Li Y, Zhou Y, Wang C, et al. Critical role of acetylation in tau-mediated neuro-degeneration and cognitive deficits. Nature Medicine. 2015;**21**:1154-1162

[146] Smet-Nocca C, Broncel M, Wieruszeski JM, Tokarski C, Hanoulle X, Leroy A, et al.

Identification of O-GlcNAc sites within peptides of the Tau protein and their impact on phosphorylation. Molecular BioSystems. 2011;**7**:1420-1429

[147] Ke YD, Suchowerska AK, van der Hoven J, De Silva DM, Wu CW, van Eersel J, et al. Lessons from tau-deficient mice. International Journal of Alzheimer's Disease. 2012;**2012**:873270

[148] Roberson ED, Scearce-Levie K, Palop JJ, Yan F, Cheng IH, Wu T, et al. Reducing endogenous tau ameliorates amyloid beta-induced deficits in an Alzheimer's disease mouse model. Science. 2007;**316**:750-754

[149] Kumar D, Ambasta RK, Kumar P. Ubiquitin biology in neurodegenerative disorders: From impairment to therapeutic strategies. Ageing Research Reviews. 2020;**61**:101078

[150] Wang W, Zhou Q, Jiang T, Li S, Ye J, Zheng J, et al. A novel small-molecule PROTAC selectively promotes tau clearance to improve cognitive functions in Alzheimer-like models. Theranostics. 2021;**11**:5279-5295

[151] Michelson D, Grundman M, Magnuson K, Fisher R, Levenson JM, Aisen P, et al. Randomized, placebo controlled trial of NPT088, A phage-derived, amyloid-targeted treatment for Alzheimer's disease. The Journal of Prevention of Alzheimer's Disease. 2019;**6**:228-231

[152] Busche MA, Hyman BT. Synergy between amyloid-beta and tau in Alzheimer's disease. Nature Neuroscience. 2020;**23**:1183-1193

[153] van der Kant R, Goldstein LSB, Ossenkoppele R. Amyloid-beta-independent regulators of tau pathology in Alzheimer disease. Nature Reviews. Neuroscience. 2020;**21**:21-35

[154] Twohig D, Nielsen HM. alpha-synuclein in the pathophysiology of Alzheimer's disease. Molecular Neurodegeneration. 2019;**14**:23

[155] Sensi SL, Granzotto A, Siotto M, Squitti R. Copper and zinc dysregulation in Alzheimer's disease. Trends in Pharmacological Sciences. 2018;**39**: 1049-1063

[156] Pradeepkiran JA, Reddy PH. Defective mitophagy in Alzheimer's disease. Ageing Research Reviews. 2020;**64**:101191

[157] Bisht K, Sharma K, Tremblay ME. Chronic stress as a risk factor for Alzheimer's disease: Roles of microglia-mediated synaptic remodeling, inflammation, and oxidative stress. Neurobiology of Stress. 2018;**9**:9-21

Chapter 6

Microglia, TREM2, and Therapeutic Methods of Alzheimer's Disease

Siwei Xu, Yaya Ji, Tianle Sha and Haoming Li

Abstract

Alzheimer's disease (AD) is one of the most common causes of dementia all around the world. It is characterized by the deposition of amyloid-β protein (Aβ) and the formation of neurofibrillary tangles (NFTs), which contribute to neuronal loss and cognitive decline. Microglia, as innate immune cells in brain, plays dual roles in the pathological process of AD. Expression in different subtypes of microglia is diverse in AD genes. Triggering receptor expressed on myeloid cells 2 (TREM2) is a transmembrane glycoprotein mainly expressed on microglia in the central nervous system (CNS). Soluble TREM2 (sTREM2), a proteolytic product of TREM2, which is abundant in the cerebrospinal fluid, shows a dynamic change in different stages and ameliorates the pathological process of AD. The interplay between the different subtypes of apolipoprotein and TREM2 is closely related to the mechanism of AD and serves as important regulatory sites. Moreover, several therapeutic strategies targeting TREM2 have shown positive outcomes during clinical trials and some novel therapies at different points are in progress. In this review, we mainly talk about the interrelationships among microglia, TREM2, and AD, and hope to give an overview of the strategies of AD.

Keywords: Alzheimer's disease (AD), microglia, TREM2, sTREM2, APOE

1. Introduction

Nowadays, Alzheimer's disease (AD) is one of the most common causes of dementia in the United States [1]. Alois Alzheimer discovered AD in 1907 and characterized AD as amyloid plaques, brain atrophy, neurofibrillary tangles, loss of neurons and synapses, and dystrophic neurites in histopathology [2].

Microglia are the resident immune cells in the CNS. They derive from erythro-myeloid progenitor cells and then migrate to the brain [3]. Developing and adult microglia demonstrate distinct morphological features as ramified or ameboid [4], which was proved by recent comprehensive transcriptomic analyses [5]. Relative analyses also demonstrate the heterogeneity, abundance, steady state in embryonic, postnatal, juvenile, and adult mouse models [6, 7]. They are also featured as self-renewing, which requires several factors such as colony-stimulating factor-1 receptor (CSF1R) and transforming growth factor β (TGF-β) [8–10]. Moreover, the murine signature of microglia in AD was present in human microglial subtypes, especially clusters 4, 5, 7, and 8. Among which, cluster 7 stands out in the consequence of its high expression of AD gene decrease in the tissue sections in both AD dementia and pathological AD [11]. This can be a diagnostic standard for AD when the frequency of cluster 7 was diminished.

Hippocampus is an elongated structure that is part of the cerebral cortex [12]. It is one of the most severely affected structures in neurodegenerative diseases like AD [13]. Hippocampus, along with its accessory structure, was suggested to be related to space [14, 15], time [16, 17], and the creation of declarative memories (memories that can evoke conscious awareness and be verbalized) [18].

Hippocampus is vulnerable to the harmfulness of diseases such as epilepsy, hypoxia, ischemia, or encephalitis [18]. The entorhinal cortex is usually the first region that demonstrates tau pathology in AD patients [18]. Somatostatin-positive interneurons are also found lost in the hippocampus of AD patients [19]. In AD patients, degenerative cholinergic neurons in the basal forebrain were proved to lead to dysfunctional cholinergic neurotransmission in regions like hippocampus [20].

2. Harmful and beneficial effects of microglia

Microglia play the role of phagocytes in the CNS, thus, maintaining the homeostasis of the brain [21]. In aging brains, microglia will cause synaptic clearance leading to forgetting *via* complement pathway [22]. In AD pathology, microglia also prove to be phagocytose synapses [23, 24]. Nevertheless, with CSF1R blockade to remove microglia in Aβ models, increased Aβ is detected [25]. Despite negative outcomes of microglia, synapse loss and behavior deficits can be avoided [26, 27]. The production of neurotoxic inflammatory cytokines and reactive oxygen species are found to be related to chronic activation of microglia [28]. However, it still remains unclear whether microglia play positive or negative roles in the process of neurodegenerative diseases.

In recent research, microglia in patients with AD show specific characteristics such as aging and upregulation of apolipoprotein E (APOE) [29]. The fat droplets appearing in microglia of aged mice suggest that the main manifestations of aging are the accumulation of fat droplets and excessive secretion of pro-inflammatory factors [30], which may be a new biological hallmark of AD. Additionally, it is not difficult to find that the branching of microglia has been reduced in aged brains, thus cutting the size of microglia's area for surveillance and leading to the harm of homeostatic functions [31–34]. One important function of microglia in AD is the phagocytosis of Aβ amyloid. For instance, microglia can mediate clearance of Aβ *via* receptors including β1 integrin in neurodegenerative diseases [35]. The acute inflammatory response can also promote phagocytosis of impaired neurons and neuronal toxic accumulation [36]. Despite the protection of microglia, prolonged inflammatory reaction will exacerbate neuronal degeneration [37]. The TAM receptor tyrosine kinases (RTKs) are a distinct family of three protein tyrosine kinases, namely Tyro3, Axl, and Mer3, which play an important role in phagocytosis and phagocytic clearance of apoptotic cells and cell membranes in the adult tissues [38, 39]. Axl and Mer play pivotal roles in macrophages like phagocytosis of apoptotic cells and negative feedback inhibition of toll-like receptor and cytokine receptor signaling. In AD mice with double knockout of Axl or Mer, the ability of microglia to phagocytize the plaque is weakened, suggesting the inhibition of TAM signal promotes plaque formation [40]. A cluster of differentiation-22 (CD22), a canonical B-cell receptor and a negative regulator of phagocytosis, is found highly expressed in microglia of aged brains, and rarely in young brains [41]. The finding suggests that the inhibition of CD22 can delay aging-related dysfunction and neurodegenerative diseases. The pellino-1 (peli1) is a ubiquitin E3 ligase, expressed in many kinds of nerve cells in the mouse brain, and with the highest expression level of microglia [42]. Similarly, Peli1 negatively regulates the ability of phagocytosis of microglia to Aβ, resulting in the inability of clearance of deposition, leading to the deterioration of AD [43].

Perineuronal nets (PNNs), with their structure remaining unknown in detail, surround the cell bodies and dendrites, and spare free space for synaptic contact [44]. In the AD mice model and human cortical tissue, PNNs are largely lost in proportion to plaque burden and depletion of microglia. Loss is prevented regardless of plaque persistence and suggests that microglia can enhance the loss of PNNs in the AD brain [45]. Besides, CD163-positive amyloid-responsive microglia are depleted in TREM2 and APOE variants in AD like TREM2 R47H and APOE4 [46].

Microglia may be detrimental to neurons in the pathological process. Recently, interleukin 3 (IL-3) from astrocytes was found to re-encode microglia, thus improving the situation of Aβ pathology [47]. Injection of IL-3 enables microglia to focus on clearing amyloid deposition and neurofibrillary tangles instead of causing extensive neuroinflammation [47]. This signaling pathway is expected to provide ideas for new drug research and development in the future and bring new drugs for the treatment of AD. A study suggested that some damaging characteristics of microglia behavior may be reversible by short-term treatment with CSF1R inhibitors [48–50]. In the mice model, removal of microglia did not improve the cognitive ability in a traumatic brain injury (TBI) [51]. Interestingly, repopulating microglia can reverse the decrease of nerve regeneration caused by brain injury and improve cognitive dysfunction in mice in an IL-6-dependant manner [51]. This study opens up a new understanding of the role of microglia in the brain injury. Remarkably, the ubiquitin ligase COP1 (also called RFWD2) is shown to dampen the neuroinflammation through inhibiting the expression of the transcription factor CCAAT/enhancer-binding protein beta (c/EBPβ), which regulates the pro-inflammatory gene of microglia [52], marking a new target for suppressing neuroinflammation in AD patients.

Disease-associated microglia (DAM), which was identified in AD patients by single-cell RNA sequencing (RNA-seq) [53], has recently become a hot topic, characterized by molecules including Iba1, Cst3, and Hexb, typically expressed in microglia. DAM also experiences downregulation of physiologically expressed genes such as P2ry12, P2ry13, Cx3cr1, CD33, and Tmem119 [9]. It is remarkable to find that DAM is identified in areas that are affected by diseases such as cortical tissue [53] and postmortem human AD brain [54]. The evidence suggests that DAM is specifically expressed in CNS pathological process, serving as an important pathological diagnostic standard. However, in the late set of neurodegenerative diseases, its role still remains unclear, which needs further investigation.

3. Physiological function on microglia of TREM2

Recent years have witnessed the central role of TREM2 as a hub in diverse pathology. TREM2 is a receptor that interacts with a variety of ligands, many of which are markers of tissue damage. TREM2 is a single-pass transmembrane protein known to regulate immune responses in peripheral macrophages through lipopolysaccharide binding and bacterial phagocytosis [55–57]. RNA-seq data were analyzed across human tissues to investigate TREM2 expression, and it has been confirmed that TREM2 is expressed physiologically in a small group of macrophages that are tissue specific [58]. In CNS, TREM2 is mostly expressed on microglia. In addition to the expression on microglia, the analysis also showed its expression in macrophages from the adrenal gland, placenta, and adipose tissue [59]. TREM2 was thought to bind a wide range of molecules [60], and the interaction with different ligands can regulate the signal intensity and direction of TREM2 in turn [61]. Downstream signals mainly consist of those arrangements; for example, DAP10 is the key to activate extracellular signal regulated-kinase (ERK) and serine/threonine protein kinase (AKT1), while in murine macrophages, DAP12 is necessary for calcium mobilization [61–63]. Functional loss of TREM2 is related to polycystic lipomembranous osteodysplasia with sclerosing leukoencephalopathy (PLOSL) [64].

Mice lacking TREM2 had defects in survival and differentiation of myeloid cells [65], as well as osteopenia and loss of microglia in CNS [62, 66]. TREM2-lacking cells may undergo a similar differential process as normal cells do despite a reduced life cycle [67]. Microglia deprived of TREM2 or expressing T66M variant demonstrated the impaired process of brain glucose metabolism and cerebral perfusion [68]. Mice TREM2 was involved in synaptic pruning through a microglia-dependent way to shape neuronal circuitry [69]. In rodents, TREM2-positive macrophages are found to be important regulators related to hair follicle stem cells [70]. Additionally, in TREM2-deficient microglia, increased autophagic vesicles can be found with defective activation of mTOR pathways [63], which partially regulate autophagy [71].

Deletion or impairment of TREM2 was proved to be detrimental to phagocytosis of lipoproteins, cellular debris, bacteria, and Aβ [68, 72, 73]. Moreover, overexpression of TREM2 in cells that are not functionally phagocytic like Chinese hamster ovary (CHO) cells showed induced phagocytosis of apoptotic cells and bacteria [57, 72].

TREM2 was also found to ameliorate neuroinflammation and neuronal apoptosis *via* PI3K/AKT signaling pathway in 5xFAD mice [74, 75]. TREM2 overexpression can also rescue cognitive barriers by reducing neuroinflammation *via* JAK/STAT/SOCS pathway [76] and the suppression of TREM2 demonstrated a defective ability to regulate the PI3K/Akt and NF-κB signaling pathways [77].

Recently, genome-wide association studies (GWAS) demonstrated a link between single-nucleotide polymorphisms (SNPs) and inflammation-related genes to increased AD risk, such as the R47H variant in TREM2 [78], which is one of the strongest genetic risk factors for AD [79]. TREM2 variant R47H, whose foundation was dysregulated peroxisome proliferator-activated receptor-γ (PPARγ)/p38MAPK signaling [80], was shown to decrease the expression of TREM2, thus deteriorating myeloid cell responses to Aβ pathology [81]. Furthermore, the R47H variants and R62H variants of TREM2 demonstrate a defective microglial transcriptional activation, implicating fully functional TREM2 seems to be the key for development of the human DAM [82].

4. TREM2 and AD

TREM2 gives protection against neurodegenerative disease. Depletion of TREM2 can induce impaired phagocytosis of the critical substrates such as APOE [83] and exacerbates tau pathology in AD [84].

4.1 TREM2 gets involved in AD pathogenesis via microglia

TREM2 is found to reduce tau seeding in neuritic plaques [85], which is essential for synapse clearance in the early stage of brain development, and TREM2-KO mice demonstrate altered sociability [69]. Moreover, TREM2 can induce microglia to gather around Aβ and restrict plaque expansion found in murine models of AD [86].

Similar conditions can also lead to exacerbation of axonal dystrophy and dendritic spine loss [87]. Another research shows that a dosage of TREM2 can reprogram the microglial response in downregulating the expression of DAM genes and ameliorating the pathological phenotype in AD mice [88]. In the absence of functional TREM2, amyloid plaque seeding increased, and microglial aggregation decreased [88]. A similar study shows that in human pluripotent stem cell (PSC), monocytes and transdifferentiated microglia-like cells, TREM2 R47H variant and loss of TREM2 on heterozygous or homozygous, display a significant decreased in phagocytosis [89]. On a recent finding, IL-4 and IL-10 enhance the phagocytosis of microglia *via* upregulation of TREM2 [90]. These findings support the hypothesis that reactive microglia and TREM2 are functionally necessary to alleviate neuronal damage. However, other studies give opposite outcomes that loss of TREM2 may be protective in AD mice [91].

Genetically, the immune cell-specific phospholipase C isoform γ2 (PLCG2), a rare coding variant, is identified [92]. Recent research has demonstrated that TREM2 can mediate phagocytosis, cell survival, lipid metabolism, and process neuronal debris through PLCG2 of microglia derived from human-induced pluripotent stem cell (iPSC) [93]. PLCG2 P552R variant has protective functions including weak-enhancing enzyme functions [94] and promoting survival functions of microglia in Plcγ2-P522R knock in mice [95]. These studies highlighted the critical role of the TREM2 pathway in AD and provided genetic evidence for the increase of TREM2 in the pathologic process of AD.

In recent years, different TREM2 ligands have been found and proposed, such as β-amyloid peptide [96] and APOE [97]. APOE-dependent molecular signature in microglia is identified in AD patients, mediating a switch from homeostatic to neurodegenerative status [98]. This can be a target in treating AD patients through restoring the homeostatic microglia.

4.2 TREM2 regulates APOE mediating AD risk

Although there is no difference in the quantity of activated microglia and reactive astrocytes between APOE4 carriers and noncarriers in the postmortem neocortex [99], relative transcriptomic studies have shown the connection between APOE and glia. Human APOE is expressed in three allelic variants, APOE2, APOE3, and APOE4, which exhibit different receptor binding properties [100]. APOE upregulation has been proved to be TREM2-dependent [101]. To some extent, TREM2 and APOE may have some special links [102], and the lack of TREM2 leads to a decrease of APOE4, while APOE3 remains unchanged [103]. Microglial plaque coverage and TREM2 are the highest in APOE3 male mice while significantly low in both APOE4 genotype and female sex [104], implicating a possible mechanism of AD between sex and APOE genotype. A reduction in plaque-associated APOE is also found in the brains of AD patients [105]. In another research, APOE3 is shown to promote the proliferation of microglia to injected Aβ, contribute to the uptake of Aβ, and improve cognition related to Aβ in preclinical models of AD [106]. Moreover, APOE was proven to stimulate different signal transduction cascades, ApoE4 > ApoE3 > ApoE2, in proportion to their AD risk [107]. This suggests that neuronal pathways may be related to the pathogenesis of AD. Human TREM2 (hTREM2) was bind to APOJ and APOE that are ligands of TREM2 under normal circumstances [73]. However, this binding is reduced in diseases or TREM2 KO mice microglia, leading to the impaired uptake of Aβ [73]. TREM2 is also an attractive target for drug regulation, but needs to be cautious because it is an important upstream mediator of microglia activation and phenotypic changes [53, 98]. In addition, single-cell transcriptomic studies pointing at microglia have shown a fascinating TREM2 ligand gal-3 that is related to neurodegenerative diseases [108]. Increased gal-3 is found in AD patients and 5xFAD mice, while decreased gal-3 shows improved cognitive ability and attenuates immune responses related to the TREM2-DAP pathway [109]. Therefore, suppressing gal-3 in the AD process may be a potential target in treatment.

5. sTREM2

sTREM2, a soluble form of TREM2, is derived from the non-proteolytic-mediated secretion of some TREM2 isoforms or due to extracellular domain of TREM2 being cleaved by different sheddases [110]. Years before the onset of dementia symptoms, sTREM2 increased in cerebrospinal fluid (CSF) of people with AD biomarker characteristics [111–114]. Recently, it is found that in preclinical AD, CSF sTREM2 changes are dynamic. In the absence of tau deposition and neurodegeneration, sTREM2 is decreased with Aβ pathology [115]. Different mutants of sTREM2 showed differences in concentration in CSF [114]. sTREM2 has a protective effect on Aβ and AD, such as reducing amyloid plaque load and restoring spatial memory [116]. Similarly, in the absence of TREM2 [105], sTREM2 enhances microglial proliferation, migration, clustering around Aβ, and contributing to the uptake and degradation of Aβ [116]. sTREM2 administration can also stimulate the expression of inflammatory cytokines and induce morphological changes of microglia such as decreased cell process and increased cell body size, thus enhancing microglial survival [117]. In TREM2 KO mice, administration of sTREM2 also showed positive feedback, like rescuing apoptosis upon colony-stimulating factor (GM-CSF) withdrawal, inducing the proliferation and cell viability of the primary microglia [118, 119], compared with WT mice [117]. These results indicate the tremendous therapeutic potential of sTREM2, but warn that pro-inflammatory activation in the brain may lead to negative functional outcomes. Under stress, sTREM2 can promote myeloid cell survival too in a manner dependent on PI3K/AKT [117, 120]. Consequently, sTREM2 can be a target for AD therapy. But it is shown less potent for sTREM2-R47H and sTREM2-R62H variants to suppress apoptosis in AD context [117]. Among the three polymorphic forms (APOE2, APOE3, and APOE4), APOE4 proved to be more related to AD [121] and high levels of sTREM2 are associated with the decrease of APOE4 [122] and slower rates of Aβ accumulation [123]. In conclusion, the changes of sTREM2 can also be a biological hallmark for AD.

6. The prospect of treatment of AD

For a long time, it is considered that Aβ accumulation is the central and initial event in the pathological process of AD. The famous amyloid cascade hypothesis thinks that the increase of Aβ levels leads to the pathological events of AD [124, 125]. Extensive clinical medicine trials of Aβ finally come to an end, and results showed that reduced Aβ load does not affect the cognitive ability of patients with AD [126, 127]. So, finding a new target rather than Aβ may be our priority. Microglia play pivotal roles in the pathological process, and interfering with their detrimental process in AD can become our next focus.

Microglia are shown to maintain the function of neurons by clearing toxic damage in the early stage of AD [128]. Consequently, interfering with the activation of microglia to lengthen the period of anti-inflammatory seems to be a therapy for AD [129]. Other anti-inflammatory cytokines such as IL-2, IL-4, and IL-33 have the potential to ameliorate AD pathology by regulating microglial activation [128], despite its results are not decisive [130].

Moreover, TREM2 is shown to be a positive target for treating AD. Recently, AL002c, an anti-human TREM2 agonistic monoclonal antibody (mAb), gives a positive outcome in 5xFAD mice expressing both the R47H variant and the common variant (CV). Prolonged administration of AL002c ameliorates filamentous plaques, causes neurodystrophy, and regulates microglial inflammation. AL002 is a derivative product of AL002c, which is modified for clinical use. AL002 is proven safe and well-tolerated in a first-in-human phase I clinical trial [131]. Overexpression of TREM2 can attenuate the pro-inflammatory effect caused by LPS, which can contribute to the increase of NO, LDH, TNF-α, IL-1b, and the activation of AKT [132]. Thus, relative experiments can be conducted in CNS.

Another way to increase TREM2 expression in microglia is by preventing ADAM10/17 family proteases from shedding extracellular domain [133, 134]. To stabilize TREM2 on the cell surface and enhance its activity, a specific mAb against TREM2 called 4D9 was screened to selectively compete for α-secretase-mediated shedding [133]. Shedding is considered to end cell-autonomous TREM2 signaling, and data show an increased phagocytic capacity of cells that express TREM2 by inhibiting ADAM proteases [135]. Combined with another research, Aβ clearance is TREM2-dependant [136], and future treatments can combine anti-Aβ antibodies with microglia-stimulating antibodies (4D9). This view opens a new door to the treatment of AD. Another study evaluated aducanumab as another antibody that may treat AD, but clinical trial results are still unsatisfactory [137]. In addition to cross-linking and activating the TREM2-DAP12 signal, 4D9 also inhibits the shedding of TREM2, resulting in the decrease of soluble TREM2 *in vitro* and the increase of total TREM2 in the brain [133]. This research may consider the role of sTREM2 in AD [91]. Since TREM2 is expressed in peripheral myeloid cells, any effects of treatment for TREM2 should be evaluated for peripheral adipose tissue in liver, lung, bone, and spleen. However, this has not been thoroughly investigated [91].

A novel property, cyclocreatine, the creatine analog, which can generate a supply chain for ATP demand regardless of the TREM2-mTOR pathway [138], is found to ameliorate autophagy, induce microglia around Aβ, and decrease neuronal dystrophy during dietary administration in 5xFAD mice [63]. Based on metabolism, this is a new era for treating AD.

Another research provides a creative angle in treating AD. It is known that meningeal lymphatic vessels drain macromolecular substances from the brain into the deep cervical lymph nodes [139], in which meningeal lymphatic serves as a channel to transport substances such as an antibody. But ablation of meningeal lymphatic vessels in 5xFAD mice can lead to a switch of microglia from homeostasis to DAM [140] and inhibit the transportation of antibodies to specific locations, thus exacerbating the cognitive ability of AD patients. It may bring unexpected clinical effects to patients with AD, if the treatment is placed in the early stage, thus enhancing the meningeal lymphatic function and combined with immunotherapy, to better play the role of meningeal lymphatic vessels.

Recently, tau pathology is the study focus. Tau hyperphosphorylation causes abnormal aggregation and neurodegeneration in AD brains [141], and protein phosphatase 2A (PP2A) has the most robust dephosphorylation activity to tau protein *in vitro* and *in vivo* [142]. A novel DEPho-sphorylation Targeting Chimaera (DEPTAC) was designed to enhance the combination of tau and PP2A-Bα, which shows high efficiency in preventing tau accumulation *in vitro* and *in vivo* [143]. Further studies showed that DEPAC significantly improved the microtubule assembly, neurite plasticity, and hippocampus-dependent learning and memory in transgenic mice [143].

7. Conclusion

Microglia play important roles in the pathological process of AD. The dual role it plays (positive or negative outcomes), its distinctive phenotype, DAM, which is specifically expressed in certain regions in AD, still needs further investigation. In most findings, TREM2 exhibits positive feedback in inhibiting detrimental factors. sTREM2, a soluble form of TREM2 in CSF, and its soluble form in CSF and sTREM2 can be biological hallmarks for diagnosis. Moreover, a close relationship between the TREM2-APOE pathway and AD demonstrates an important pathological feature. A new therapeutic method based on TREM2 to manipulate the function of microglia is currently being tested. Although there are still numerous obstacles ahead to treating AD, it is expected that this field will move closer to understanding the influence of microglia regulation in AD, which is a breakthrough result for patients. Most therapeutic treatments targeting Aβ do not get expected feedback. Thus, genetic evidence and metabolic mechanism related to AD should be more explored in future studies.

Author details

Siwei Xu, Yaya Ji, Tianle Sha and Haoming Li*
Department of Human Anatomy, Medical School of Nantong University, Nantong, China

*Address all correspondence to: lihaoming@ntu.edu.cn

References

[1] James BD, Leurgans SE, Hebert LE, Scherr PA, Yaffe K, Bennett DA. Contribution of Alzheimer disease to mortality in the United States. Neurology. 2014;**82**(12):1045-1050. DOI: 10.1212/WNL.00000000000 00240

[2] Alzheimer, AJAzp. Uber eine eigenartige Erkrankung der Hirnrinde. Allgemeine Zeitschrife Psychiatrie.1907; 64: 146-148.

[3] Ginhoux F, Lim S, Hoeffel G, Low D, Huber T. Origin and differentiation of microglia. Frontiers in Cellular Neuroscience. 2013;**7**:45. DOI: 10.3389/fncel.2013.00045

[4] Lawson LJ, Perry VH, Gordon S. Turnover of resident microglia in the normal adult mouse brain. Neuro science. 1992;**48**(2):405-415. DOI: 10.1016/0306-4522(92)90500-2

[5] Matcovitch-Natan O, Winter DR, Giladi A, Vargas Aguilar S, Spinrad A, Sarrazin S, et al. Microglia development follows a stepwise program to regulate brain homeostasis. Science. 2016;**353** (6301):aad8670. DOI: 10.1126/science.aad8670

[6] Varol D, Mildner A, Blank T, Shemer A, Barashi N, Yona S, et al. Dicer deficiency differentially impacts microglia of the developing and adult brain. Immunity. 2017;**46**(6):1030-1044 e8. DOI: 10.1016/j.immuni.2017.05.003

[7] Masuda T, Sankowski R, Staszewski O, Bottcher C, Amann L, Sagar, et al. Spatial and temporal heterogeneity of mouse and human microglia at single-cell resolution. Nature. 2019;**566**(7744):388-392. DOI: 10.1038/s41586-019-0924-x

[8] Ginhoux F, Greter M, Leboeuf M, Nandi S, See P, Gokhan S, et al. Fate mapping analysis reveals that adult microglia derive from primitive macrophages. Science. 2010;**330**(6005): 841-845. DOI: 10.1126/science.1194637

[9] Butovsky O, Jedrychowski MP, Moore CS, Cialic R, Lanser AJ, Gabriely G, et al. Identification of a unique TGF-beta-dependent molecular and functional signature in microglia. Nature Neuroscience. 2014;**17**(1):131-143. DOI: 10.1038/nn.3599

[10] Elmore MR, Najafi AR, Koike MA, Dagher NN, Spangenberg EE, Rice RA, et al. Colony-stimulating factor 1 receptor signaling is necessary for microglia viability, unmasking a microglia progenitor cell in the adult brain. Neuron. 2014;**82**(2):380-397. DOI: 10.1016/j.neuron.2014.02.040

[11] Olah M, Menon V, Habib N, Taga MF, Ma Y, Yung CJ, et al. Single cell RNA sequencing of human microglia uncovers a subset associated with Alzheimer's disease. Nature Communications. 2020;**11**(1):6129. DOI: 10.1038/s41467-020-19737-2

[12] Gilbert PE, Brushfield AM. The role of the CA3 hippocampal subregion in spatial memory: A process oriented behavioral assessment. Progress in Neuro-Psychopharmacology & Biological Psychiatry. 2009;**33**(5):774-781. DOI: 10.1016/j.pnpbp.2009.03.037

[13] Frisoni GB, Fox NC, Jack CR Jr, Scheltens P, Thompson PM. The clinical use of structural MRI in Alzheimer disease. Nature Reviews. Neurology. 2010;**6**(2):67-77. DOI: 10.1038/nrneurol.2009.215

[14] Eckardt, Michael. The Hippocampus as a Cognitive Map. The Journal of Nervous and Mental Disease.1980; 168: 191-192.DOI: 10.1097/00005053-198003000-00018.

[15] Moser EI, Roudi Y, Witter MP, Kentros C, Bonhoeffer T, Moser MB. Grid cells and cortical representation. Nature Reviews. Neuroscience. 2014;**15**(7):466-481. DOI: 10.1038/nrn3766

[16] Eichenbaum H. Time cells in the hippocampus: A new dimension for mapping memories. Nature Reviews. Neuroscience. 2014;**15**(11):732-744. DOI: 10.1038/nrn3827

[17] Eichenbaum H. On the integration of space, time, and memory. Neuron. 2017;**95**(5):1007-1018. DOI: 10.1016/j.neuron.2017.06.036

[18] Knierim JJ. The hippocampus. Current Biology. 2015;**25**(23):R1116-R1121. DOI: 10.1016/j.cub.2015.10.049

[19] Chan-Palay V. Somatostatin immunoreactive neurons in the human hippocampus and cortex shown by immunogold/silver intensification on vibratome sections: Coexistence with neuropeptide Y neurons, and effects in Alzheimer-type dementia. The Journal of Comparative Neurology. 1987;**260**(2):201-223. DOI: 10.1002/cne.902600205

[20] Perry E, Walker M, Grace J, Perry R. Acetylcholine in mind: A neurotransmitter correlate of consciousness? Trends in Neurosciences. 1999;**22**(6):273-280. DOI: 10.1016/s0166-2236(98)01361-7

[21] Nayak D, Roth TL, McGavern DB. Microglia development and function. Annual Review of Immunology. 2014;**32**:367-402. DOI: 10.1146/annurev-immunol-032713-120240

[22] Wang C, Yue H, Hu Z, Shen Y, Ma J, Li J, et al. Microglia mediate forgetting via complement-dependent synaptic elimination. Science. 2020;**367**(6478):688-694. DOI: 10.1126/science.aaz2288

[23] Shi Q, Chowdhury S, Ma R, Le KX, Hong S, Caldarone BJ, et al. Complement C3 deficiency protects against neurodegeneration in aged plaque-rich APP/PS1 mice. Science Translational Medicine. 2017;**9**(392):eaaf6295. DOI: 10.1126/scitranslmed.aaf6295

[24] Hong S, Beja-Glasser VF, Nfonoyim BM, Frouin A, Li S, Ramakrishnan S, et al. Complement and microglia mediate early synapse loss in Alzheimer mouse models. Science. 2016;**352**(6286):712-716. DOI: 10.1126/science.aad8373

[25] Dagher NN, Najafi AR, Kayala KM, Elmore MR, White TE, Medeiros R, et al. Colony-stimulating factor 1 receptor inhibition prevents microglial plaque association and improves cognition in 3xTg-AD mice. Journal of Neuroinflammation. 2015;**12**:139. DOI: 10.1186/s12974-015-0366-9

[26] Olmos-Alonso A, Schetters ST, Sri S, Askew K, Mancuso R, Vargas-Caballero M, et al. Pharmacological targeting of CSF1R inhibits microglial proliferation and prevents the progression of Alzheimer's-like pathology. Brain. 2016;**139**(Pt 3):891-907. DOI: 10.1093/brain/awv379

[27] Spangenberg EE, Lee RJ, Najafi AR, Rice RA, Elmore MR, Blurton-Jones M, et al. Eliminating microglia in Alzheimer's mice prevents neuronal loss without modulating amyloid-beta pathology. Brain. 2016;**139**(Pt 4):1265-1281. DOI: 10.1093/brain/aww016

[28] Block ML, Zecca L, Hong JS. Microglia-mediated neurotoxicity: Uncovering the molecular mechanisms. Nature Reviews. Neuroscience. 2007;**8**(1):57-69. DOI: 10.1038/nrn2038

[29] Srinivasan K, Friedman BA, Etxeberria A, Huntley MA, van der Brug MP, Foreman O, et al. Alzheimer's patient microglia exhibit enhanced aging and unique transcriptional

activation. Cell Reports. 2020;**31**(13): 107843. DOI: 10.1016/j.celrep.2020.107843

[30] Marschallinger J, Iram T, Zardeneta M, Lee SE, Lehallier B, Haney MS, et al. Lipid-droplet-accumulating microglia represent a dysfunctional and proinflammatory state in the aging brain. Nature Neuroscience. 2020;**23**(2):194-208. DOI: 10.1038/s41593-019-0566-1

[31] Spittau B. Aging microglia-phenotypes, functions and implications for age-related neurodegenerative diseases. Frontiers in Aging Neuro science. 2017;**9**:194. DOI: 10.3389/fnagi.2017.00194

[32] Davies DS, Ma J, Jegathees T, Goldsbury C. Microglia show altered morphology and reduced arborization in human brain during aging and Alzheimer's disease. Brain Pathology. 2017;**27**(6):795-808. DOI: 10.1111/bpa.12456

[33] Rawji KS, Mishra MK, Michaels NJ, Rivest S, Stys PK, Yong VW. Immunosenescence of microglia and macrophages: Impact on the ageing central nervous system. Brain. 2016;**139**(Pt 3):653-661. DOI: 10.1093/brain/awv395

[34] Bisht K, Sharma KP, Lecours C, Sanchez MG, El Hajj H, Milior G, et al. Dark microglia: A new phenotype predominantly associated with pathological states. Glia. 2016;**64**(5): 826-839. DOI: 10.1002/glia.22966

[35] Koenigsknecht J, Landreth G. Microglial phagocytosis of fibrillar beta-amyloid through a beta1 integrin-dependent mechanism. The Journal of Neuroscience. 2004;**24**(44):9838-9846. DOI: 10.1523/JNEUROSCI.2557-04.2004

[36] Neher JJ, Neniskyte U, Zhao JW, Bal-Price A, Tolkovsky AM, Brown GC. Inhibition of microglial phagocytosis is sufficient to prevent inflammatory neuronal death. Journal of Immunology. 2011;**186**(8):4973-4983. DOI: 10.4049/jimmunol.1003600

[37] Shabab T, Khanabdali R, Moghadamtousi SZ, Kadir HA, Mohan G. Neuroinflammation pathways: A general review. The International Journal of Neuroscience. 2017;**127**(7):624-633. DOI: 10.1080/00207454.2016.1212854

[38] Lemke G, Rothlin CV. Immunobiology of the TAM receptors. Nature Reviews. Immunology. 2008;**8**(5):327-336. DOI: 10.1038/nri2303

[39] Lemke G, Burstyn-Cohen T. TAM receptors and the clearance of apoptotic cells. Annals of the New York Academy of Sciences. 2010;**1209**:23-29. DOI: 10.1111/j.1749-6632.2010.05744.x

[40] Huang Y, Happonen KE, Burrola PG, O'Connor C, Hah N, Huang L, et al. Microglia use TAM receptors to detect and engulf amyloid beta plaques. Nature Immunology. 2021;**22**(5):586-594. DOI: 10.1038/s41590-021-00913-5

[41] Pluvinage JV, Haney MS, Smith BAH, Sun J, Iram T, Bonanno L, et al. CD22 blockade restores homeostatic microglial phagocytosis in ageing brains. Nature. 2019;**568**(7751):187-192. DOI: 10.1038/s41586-019-1088-4

[42] Xiao Y, Jin J, Chang M, Chang JH, Hu H, Zhou X, et al. Peli1 promotes microglia-mediated CNS inflammation by regulating Traf3 degradation. Nature Medicine. 2013;**19**(5):595-602. DOI: 10.1038/nm.3111

[43] Xu J, Yu T, Pietronigro EC, Yuan J, Arioli J, Pei Y, et al. Peli1 impairs microglial Abeta phagocytosis through promoting C/EBPbeta degradation. PLoS Biology. 2020;**18**(10):e3000837. DOI: 10.1371/journal.pbio.3000837

[44] Matthews RT, Kelly GM, Zerillo CA, Gray G, Tiemeyer M, Hockfield S. Aggrecan glycoforms contribute to the molecular heterogeneity of perineuronal nets. The Journal of Neuroscience. 2002;**22**(17):7536-7547

[45] Crapser JD, Spangenberg EE, Barahona RA, Arreola MA, Hohsfield LA, Green KN. Microglia facilitate loss of perineuronal nets in the Alzheimer's disease brain. eBioMedicine. 2020;**58**:102919. DOI: 10.1016/j.ebiom.2020.102919

[46] Nguyen AT, Wang K, Hu G, Wang X, Miao Z, Azevedo JA, et al. APOE and TREM2 regulate amyloid-responsive microglia in Alzheimer's disease. Acta Neuropathologica. 2020;**140**(4):477-493. DOI: 10.1007/s00401-020-02200-3

[47] McAlpine CS, Park J, Griciuc A, Kim E, Choi SH, Iwamoto Y, et al. Astrocytic interleukin-3 programs microglia and limits Alzheimer's disease. Nature. 2021;**595**(7869):701-706. DOI: 10.1038/s41586-021-03734-6

[48] Rice RA, Spangenberg EE, Yamate-Morgan H, Lee RJ, Arora RP, Hernandez MX, et al. Elimination of microglia improves functional outcomes following extensive neuronal loss in the hippocampus. The Journal of Neuroscience. 2015;**35**(27):9977-9989. DOI: 10.1523/JNEUROSCI.0336-15.2015

[49] Rice RA, Pham J, Lee RJ, Najafi AR, West BL, Green KN. Microglial repopulation resolves inflammation and promotes brain recovery after injury. Glia. 2017;**65**(6):931-944. DOI: 10.1002/glia.23135

[50] Weber MD, McKim DB, Niraula A, Witcher KG, Yin W, Sobol CG, et al. The influence of microglial elimination and repopulation on stress sensitization induced by repeated social defeat. Biological Psychiatry. 2019;**85**(8):667-678. DOI: 10.1016/j.biopsych.2018.10.009

[51] Willis EF, MacDonald KPA, Nguyen QH, Garrido AL, Gillespie ER, Harley SBR, et al. Repopulating microglia promote brain repair in an IL-6-dependent manner. Cell. 2020;**180**(5):833-846 e16. DOI: 10.1016/j.cell.2020.02.013

[52] Ndoja A, Reja R, Lee SH, Webster JD, Ngu H, Rose CM, et al. Ubiquitin ligase COP1 suppresses neuroinflammation by degrading c/EBPbeta in microglia. Cell. 2020;**182**(5):1156-1169 e12. DOI: 10.1016/j.cell.2020.07.011

[53] Keren-Shaul H, Spinrad A, Weiner A, Matcovitch-Natan O, Dvir-Szternfeld R, Ulland TK, et al. A unique microglia type associated with restricting development of Alzheimer's disease. Cell. 2017;**169**(7):1276-1290 e17. DOI: 10.1016/j.cell.2017.05.018

[54] Friedman BA, Srinivasan K, Ayalon G, Meilandt WJ, Lin H, Huntley MA, et al. Diverse brain myeloid expression profiles reveal distinct microglial activation states and aspects of Alzheimer's disease not evident in mouse models. Cell Reports. 2018;**22**(3):832-847. DOI: 10.1016/j.celrep.2017.12.066

[55] Turnbull IR, Gilfillan S, Cella M, Aoshi T, Miller M, Piccio L, et al. Cutting edge: TREM-2 attenuates macrophage activation. Journal of Immunology. 2006;**177**(6):3520-3524. DOI: 10.4049/jimmunol.177.6.3520

[56] Hamerman JA, Jarjoura JR, Humphrey MB, Nakamura MC, Seaman WE, Lanier LL. Cutting edge: Inhibition of TLR and FcR responses in macrophages by triggering receptor expressed on myeloid cells (TREM)-2 and DAP12. Journal of Immunology. 2006;**177**(4):2051-2055. DOI: 10.4049/jimmunol.177.4.2051

[57] N'Diaye EN, Branda CS, Branda SS, Nevarez L, Colonna M, Lowell C, et al. TREM-2 (triggering receptor expressed on myeloid cells 2) is a phagocytic receptor for bacteria. The Journal of Cell Biology. 2009;**184**(2):215-223. DOI: 10.1083/jcb.200808080

[58] Han X, Zhou Z, Fei L, Sun H, Wang R, Chen Y, et al. Construction of a human cell landscape at single-cell level. Nature. 2020;**581**(7808):303-309. DOI: 10.1038/s41586-020-2157-4

[59] Jaitin DA, Adlung L, Thaiss CA, Weiner A, Li B, Descamps H, et al. Lipid-associated macrophages control metabolic homeostasis in a Trem2-dependent manner. Cell. 2019;**178**(3): 686-698 e14. DOI: 10.1016/j.cell.2019.05.054

[60] Kober DL, Brett TJ. TREM2-ligand interactions in health and disease. Journal of Molecular Biology. 2017;**429**(11):1607-1629. DOI: 10.1016/j.jmb.2017.04.004

[61] Peng Q, Malhotra S, Torchia JA, Kerr WG, Coggeshall KM, Humphrey MB. TREM2- and DAP12-dependent activation of PI3K requires DAP10 and is inhibited by SHIP1. Science Signaling. 2010;**3**(122):ra38. DOI: 10.1126/scisignal.2000500

[62] Otero K, Shinohara M, Zhao H, Cella M, Gilfillan S, Colucci A, et al. TREM2 and beta-catenin regulate bone homeostasis by controlling the rate of osteoclastogenesis. Journal of Immunology. 2012;**188**(6):2612-2621. DOI: 10.4049/jimmunol.1102836

[63] Ulland TK, Song WM, Huang SC, Ulrich JD, Sergushichev A, Beatty WL, et al. TREM2 maintains microglial metabolic fitness in Alzheimer's disease. Cell. 2017;**170**(4):649-663 e13. DOI: 10.1016/j.cell.2017.07.023

[64] Klunemann HH, Ridha BH, Magy L, Wherrett JR, Hemelsoet DM, Keen RW, et al. The genetic causes of basal ganglia calcification, dementia, and bone cysts: DAP12 and TREM2. Neurology. 2005;**64**(9):1502-1507. DOI: 10.1212/01.WNL.0000160304.00003.CA

[65] Otero K, Turnbull IR, Poliani PL, Vermi W, Cerutti E, Aoshi T, et al. Macrophage colony-stimulating factor induces the proliferation and survival of macrophages via a pathway involving DAP12 and beta-catenin. Nature Immunology. 2009;**10**(7):734-743. DOI: 10.1038/ni.1744

[66] Poliani PL, Wang Y, Fontana E, Robinette ML, Yamanishi Y, Gilfillan S, et al. TREM2 sustains microglial expansion during aging and response to demyelination. The Journal of Clinical Investigation. 2015;**125**(5):2161-2170. DOI: 10.1172/JCI77983

[67] Deczkowska A, Keren-Shaul H, Weiner A, Colonna M, Schwartz M, Amit I. Disease-associated microglia: A universal immune sensor of neurodegeneration. Cell. 2018;**173**(5): 1073-1081. DOI: 10.1016/j.cell.2018.05.003

[68] Kleinberger G, Brendel M, Mracsko E, Wefers B, Groeneweg L, Xiang X, et al. The FTD-like syndrome causing TREM2 T66M mutation impairs microglia function, brain perfusion, and glucose metabolism. EMBO Journal. 2017;**36**(13):1837-1853. DOI: 10.15252/embj.201796516

[69] Filipello F, Morini R, Corradini I, Zerbi V, Canzi A, Michalski B, et al. The microglial innate immune receptor TREM2 is required for synapse elimination and normal brain connectivity. Immunity. 2018;**48**(5): 979-991 e8. DOI: 10.1016/j.immuni.2018.04.016

[70] Wang ECE, Dai Z, Ferrante AW, Drake CG, Christiano AM. A subset of TREM2(+) dermal macrophages secretes oncostatin M to maintain hair

follicle stem cell quiescence and inhibit hair growth. Cell Stem Cell. 2019;**24**(4):654-669 e6. DOI: 10.1016/j.stem.2019.01.011

[71] Saxton RA, Sabatini DM. mTOR signaling in growth, metabolism, and disease. Cell. 2017;**168**(6):960-976. DOI: 10.1016/j.cell.2017.02.004

[72] Takahashi K, Rochford CD, Neumann H. Clearance of apoptotic neurons without inflammation by microglial triggering receptor expressed on myeloid cells-2. The Journal of Experimental Medicine. 2005;**201**(4): 647-657. DOI: 10.1084/jem.20041611

[73] Yeh FL, Wang Y, Tom I, Gonzalez LC, Sheng M. TREM2 binds to apolipoproteins, including APOE and CLU/APOJ, and thereby facilitates uptake of amyloid-beta by microglia. Neuron. 2016;**91**(2):328-340. DOI: 10.1016/j.neuron.2016.06.015

[74] Wang Y, Lin Y, Wang L, Zhan H, Luo X, Zeng Y, et al. TREM2 ameliorates neuroinflammatory response and cognitive impairment via PI3K/AKT/FoxO3a signaling pathway in Alzheimer's disease mice. Aging (Albany NY). 2020;**12**(20):20862, 10.18632/aging.104104-20879

[75] Chen S, Peng J, Sherchan P, Ma Y, Xiang S, Yan F, et al. TREM2 activation attenuates neuroinflammation and neuronal apoptosis via PI3K/Akt pathway after intracerebral hemorrhage in mice. Journal of Neuroinflammation. 2020;**17**(1):168. DOI: 10.1186/s12974-020-01853-x

[76] Ruganzu JB, Zheng Q, Wu X, He Y, Peng X, Jin H, et al. TREM2 overexpression rescues cognitive deficits in APP/PS1 transgenic mice by reducing neuroinflammation via the JAK/STAT/SOCS signaling pathway. Experimental Neurology. 2021;**336**:113506. DOI: 10.1016/j.expneurol.2020.113506

[77] Xu H, Mu S, Qin W. Microglia TREM2 is required for electroacupuncture to attenuate neuroinflammation in focal cerebral ischemia/reperfusion rats. Biochemical and Biophysical Research Communications. 2018;**503**(4):3225-3234. DOI: 10.1016/j.bbrc.2018.08.130

[78] Karch CM, Goate AM. Alzheimer's disease risk genes and mechanisms of disease pathogenesis. Biological Psychiatry. 2015;**77**(1):43-51. DOI: 10.1016/j.biopsych.2014.05.006

[79] Guerreiro R, Wojtas A, Bras J, Carrasquillo M, Rogaeva E, Majounie E, et al. TREM2 variants in Alzheimer's disease. The New England Journal of Medicine. 2013;**368**(2):117-127. DOI: 10.1056/NEJMoa1211851

[80] Piers TM, Cosker K, Mallach A, Johnson GT, Guerreiro R, Hardy J, et al. A locked immunometabolic switch underlies TREM2 R47H loss of function in human iPSC-derived microglia. The FASEB Journal. 2020;**34**(2):2436-2450. DOI: 10.1096/fj.201902447R

[81] Cheng-Hathaway PJ, Reed-Geaghan EG, Jay TR, Casali BT, Bemiller SM, Puntambekar SS, et al. The Trem2 R47H variant confers loss-of-function-like phenotypes in Alzheimer's disease. Molecular Neurodegeneration. 2018;**13**(1):29. DOI: 10.1186/s13024-018-0262-8

[82] Zhou Y, Song WM, Andhey PS, Swain A, Levy T, Miller KR, et al. Human and mouse single-nucleus transcriptomics reveal TREM2-dependent and TREM2-independent cellular responses in Alzheimer's disease. Nature Medicine. 2020;**26**(1): 131-142. DOI: 10.1038/s41591-019-0695-9

[83] McQuade A, Kang YJ, Hasselmann J, Jairaman A, Sotelo A, Coburn M, et al. Gene expression and functional deficits

underlie TREM2-knockout microglia responses in human models of Alzheimer's disease. Nature Communications. 2020;**11**(1):5370. DOI: 10.1038/s41467-020-19227-5

[84] Bemiller SM, McCray TJ, Allan K, Formica SV, Xu G, Wilson G, et al. TREM2 deficiency exacerbates tau pathology through dysregulated kinase signaling in a mouse model of tauopathy. Molecular Neuro degeneration. 2017;**12**(1):74. DOI: 10.1186/s13024-017-0216-6

[85] Leyns CEG, Gratuze M, Narasimhan S, Jain N, Koscal LJ, Jiang H, et al. TREM2 function impedes tau seeding in neuritic plaques. Nature Neuroscience. 2019;**22**(8):1217-1222. DOI: 10.1038/s41593-019-0433-0

[86] Wang Y, Ulland TK, Ulrich JD, Song W, Tzaferis JA, Hole JT, et al. TREM2-mediated early microglial response limits diffusion and toxicity of amyloid plaques. The Journal of Experimental Medicine. 2016;**213**(5): 667-675. DOI: 10.1084/jem.20151948

[87] Meilandt WJ, Ngu H, Gogineni A, Lalehzadeh G, Lee SH, Srinivasan K, et al. Trem2 deletion reduces late-stage amyloid plaque accumulation, elevates the Abeta42:Abeta40 ratio, and exacerbates axonal dystrophy and dendritic spine loss in the PS2APP Alzheimer's mouse model. The Journal of Neuroscience. 2020;**40**(9):1956-1974. DOI: 10.1523/JNEUROSCI.1871-19.2019

[88] Lee CYD, Daggett A, Gu X, Jiang LL, Langfelder P, Li X, et al. Elevated TREM2 gene dosage reprograms microglia responsivity and ameliorates pathological phenotypes in Alzheimer's disease models. Neuron. 2018;**97**(5):1032-1048 e5. DOI: 10.1016/j.neuron.2018.02.002

[89] Claes C, Van Den Daele J, Boon R, Schouteden S, Colombo A, Monasor LS, et al. Human stem cell-derived monocytes and microglia-like cells reveal impaired amyloid plaque clearance upon heterozygous or homozygous loss of TREM2. Alzheimer's & Dementia. 2019;**15**(3):453-464. DOI: 10.1016/j.jalz.2018.09.006

[90] Yi S, Jiang X, Tang X, Li Y, Xiao C, Zhang J, et al. IL-4 and IL-10 promotes phagocytic activity of microglia by up-regulation of TREM2. Cytotechnology. 2020;**72**(4):589-602. DOI: 10.1007/s10616-020-00409-4

[91] Lewcock JW, Schlepckow K, Di Paolo G, Tahirovic S, Monroe KM, Haass C. Emerging microglia biology defines novel therapeutic approaches for Alzheimer's disease. Neuron. 2020;**108**(5):801-821. DOI: 10.1016/j.neuron.2020.09.029

[92] Sims R, van der Lee SJ, Naj AC, Bellenguez C, Badarinarayan N, Jakobsdottir J, et al. Rare coding variants in PLCG2, ABI3, and TREM2 implicate microglial-mediated innate immunity in Alzheimer's disease. Nature Genetics. 2017;**49**(9):1373-1384. DOI: 10.1038/ng.3916

[93] Andreone BJ, Przybyla L, Llapashtica C, Rana A, Davis SS, van Lengerich B, et al. Alzheimer's-associated PLCgamma2 is a signaling node required for both TREM2 function and the inflammatory response in human microglia. Nature Neuroscience. 2020;**23**(8):927-938. DOI: 10.1038/s41593-020-0650-6

[94] Magno L, Lessard CB, Martins M, Lang V, Cruz P, Asi Y, et al. Alzheimer's disease phospholipase C-gamma-2 (PLCG2) protective variant is a functional hypermorph. Alzheimer's Research & Therapy. 2019;**11**(1):16. DOI: 10.1186/s13195-019-0469-0

[95] Takalo M, Wittrahm R, Wefers B, Parhizkar S, Jokivarsi K, Kuulasmaa T, et al. The Alzheimer's disease-associated protective Plcgamma2-P522R variant

promotes immune functions. Molecular Neurodegeneration. 2020;**15**(1):52. DOI: 10.1186/s13024-020-00402-7

[96] Zhao Y, Wu X, Li X, Jiang LL, Gui X, Liu Y, et al. TREM2 Is a receptor for beta-amyloid that mediates microglial function. Neuron. 2018;**97**(5):1023-1031 e7. DOI: 10.1016/j.neuron.2018.01.031

[97] Atagi Y, Liu CC, Painter MM, Chen XF, Verbeeck C, Zheng H, et al. Apolipoprotein E is a ligand for triggering receptor expressed on myeloid cells 2 (TREM2). The Journal of Biological Chemistry. 2015;**290**(43):26043-26050. DOI: 10.1074/jbc.M115.679043

[98] Krasemann S, Madore C, Cialic R, Baufeld C, Calcagno N, El Fatimy R, et al. The TREM2-APOE pathway drives the transcriptional phenotype of dysfunctional microglia in neurodegenerative diseases. Immunity. 2017;**47**(3):566-581 e9. DOI: 10.1016/j.immuni.2017.08.008

[99] Serrano-Pozo A, Betensky RA, Frosch MP, Hyman BT. Plaque-associated local toxicity increases over the clinical course of Alzheimer disease. The American Journal of Pathology. 2016;**186**(2):375-384. DOI: 10.1016/j.ajpath.2015.10.010

[100] Calandra S, Tarugi P, Speedy HE, Dean AF, Bertolini S, Shoulders CC. Mechanisms and genetic determinants regulating sterol absorption, circulating LDL levels, and sterol elimination: Implications for classification and disease risk. Journal of Lipid Research. 2011;**52**(11):1885-1926. DOI: 10.1194/jlr.R017855

[101] Nugent AA, Lin K, van Lengerich B, Lianoglou S, Przybyla L, Davis SS, et al. TREM2 regulates microglial cholesterol metabolism upon chronic phagocytic challenge. Neuron. 2020;**105**(5):837-854 e9. DOI: 10.1016/j.neuron.2019.12.007

[102] Damisah EC, Rai A, Grutzendler J. TREM2: Modulator of lipid metabolism in microglia. Neuron. 2020;**105**(5):759-761. DOI: 10.1016/j.neuron.2020.02.008

[103] Fitz NF, Wolfe CM, Playso BE, Biedrzycki RJ, Lu Y, Nam KN, et al. Trem2 deficiency differentially affects phenotype and transcriptome of human APOE3 and APOE4 mice. Molecular Neurodegeneration. 2020;**15**(1):41. DOI: 10.1186/s13024-020-00394-4

[104] Stephen TL, Cacciottolo M, Balu D, Morgan TE, LaDu MJ, Finch CE, et al. APOE genotype and sex affect microglial interactions with plaques in Alzheimer's disease mice. Acta Neuropathologica Communications. 2019;**7**(1):82. DOI: 10.1186/s40478-019-0729-z

[105] Parhizkar S, Arzberger T, Brendel M, Kleinberger G, Deussing M, Focke C, et al. Loss of TREM2 function increases amyloid seeding but reduces plaque-associated ApoE. Nature Neuroscience. 2019;**22**(2):191-204. DOI: 10.1038/s41593-018-0296-9

[106] Fitz NF, Nam KN, Wolfe CM, Letronne F, Playso BE, Iordanova BE, et al. Phospholipids of APOE lipoproteins activate microglia in an isoform-specific manner in preclinical models of Alzheimer's disease. Nature Communications. 2021;**12**(1):3416. DOI: 10.1038/s41467-021-23762-0

[107] Huang YA, Zhou B, Nabet AM, Wernig M, Sudhof TC. Differential signaling mediated by ApoE2, ApoE3, and ApoE4 in human neurons parallels Alzheimer's disease risk. The Journal of Neuroscience. 2019;**39**(37):7408-7427. DOI: 10.1523/JNEUROSCI.2994-18.2019

[108] Butovsky O, Weiner HL. Microglial signatures and their role in health and disease. Nature Reviews. Neuroscience. 2018;**19**(10):622-635. DOI: 10.1038/s41583-018-0057-5

[109] Boza-Serrano A, Ruiz R, Sanchez-Varo R, Garcia-Revilla J, Yang Y, Jimenez-Ferrer I, et al. Galectin-3, a novel endogenous TREM2 ligand, detrimentally regulates inflammatory response in Alzheimer's disease. Acta Neuropathologica. 2019;**138**(2):251-273. DOI: 10.1007/s00401-019-02013-z

[110] Yang J, Fu Z, Zhang X, Xiong M, Meng L, Zhang Z. TREM2 ectodomain and its soluble form in Alzheimer's disease. Journal of Neuroinflammation. 2020;**17**(1):204. DOI: 10.1186/s12974-020-01878-2

[111] Heslegrave A, Heywood W, Paterson R, Magdalinou N, Svensson J, Johansson P, et al. Increased cerebrospinal fluid soluble TREM2 concentration in Alzheimer's disease. Molecular Neurodegeneration. 2016;**11**:3. DOI: 10.1186/s13024-016-0071-x

[112] Piccio L, Deming Y, Del-Aguila JL, Ghezzi L, Holtzman DM, Fagan AM, et al. Cerebrospinal fluid soluble TREM2 is higher in Alzheimer disease and associated with mutation status. Acta Neuropathologica. 2016;**131**(6):925-933. DOI: 10.1007/s00401-016-1533-5

[113] Suarez-Calvet M, Araque Caballero MA, Kleinberger G, Bateman RJ, Fagan AM, Morris JC, et al. Early changes in CSF sTREM2 in dominantly inherited Alzheimer's disease occur after amyloid deposition and neuronal injury. Science Translational Medicine. 2016;**8**(369):369ra178. DOI: 10.1126/scitranslmed.aag1767

[114] Suarez-Calvet M, Morenas-Rodriguez E, Kleinberger G, Schlepckow K, Araque Caballero MA, Franzmeier N, et al. Early increase of CSF sTREM2 in Alzheimer's disease is associated with tau related-neurodegeneration but not with amyloid-beta pathology. Molecular Neurodegeneration. 2019;**14**(1):1. DOI: 10.1186/s13024-018-0301-5

[115] Ma LZ, Tan L, Bi YL, Shen XN, Xu W, Ma YH, et al. Dynamic changes of CSF sTREM2 in preclinical Alzheimer's disease: The CABLE study. Molecular Neurodegeneration. 2020;**15**(1):25. DOI: 10.1186/s13024-020-00374-8

[116] Zhong L, Xu Y, Zhuo R, Wang T, Wang K, Huang R, et al. Author correction: Soluble TREM2 ameliorates pathological phenotypes by modulating microglial functions in an Alzheimer's disease model. Nature Communications. 2019;**10**(1):2923. DOI: 10.1038/s41467-019-10950-2

[117] Zhong L, Chen XF, Wang T, Wang Z, Liao C, Wang Z, et al. Soluble TREM2 induces inflammatory responses and enhances microglial survival. The Journal of Experimental Medicine. 2017;**214**(3):597-607. DOI: 10.1084/jem.20160844

[118] Lee SC, Liu W, Brosnan CF, Dickson DW. GM-CSF promotes proliferation of human fetal and adult microglia in primary cultures. Glia. 1994;**12**(4):309-318. DOI: 10.1002/glia.440120407

[119] Tomozawa Y, Inoue T, Takahashi M, Adachi M, Satoh M. Apoptosis of cultured microglia by the deprivation of macrophage colony-stimulating factor. Neuroscience Research. 1996;**25**(1):7-15. DOI: 10.1016/0168-0102(96)01021-8

[120] Wu K, Byers DE, Jin X, Agapov E, Alexander-Brett J, Patel AC, et al. TREM-2 promotes macrophage survival and lung disease after respiratory viral infection. The Journal of Experimental Medicine. 2015;**212**(5):681-697. DOI: 10.1084/jem.20141732

[121] Strittmatter WJ, Roses AD. Apolipoprotein E and Alzheimer's

disease. Annual Review of Neuroscience. 1996;**19**:53-77. DOI: 10.1146/annurev.ne.19.030196.000413

[122] Franzmeier N, Suarez-Calvet M, Frontzkowski L, Moore A, Hohman TJ, Morenas-Rodriguez E, et al. Higher CSF sTREM2 attenuates ApoE4-related risk for cognitive decline and neuro degeneration. Molecular Neurodegeneration. 2020;**15**(1):57. DOI: 10.1186/s13024-020-00407-2

[123] Ewers M, Biechele G, Suarez-Calvet M, Sacher C, Blume T, Morenas-Rodriguez E, et al. Higher CSF sTREM2 and microglia activation are associated with slower rates of beta-amyloid accumulation. EMBO Molecular Medicine. 2020;**12**(9):e12308. DOI: 10.15252/emmm.202012308

[124] Hardy J, Allsop D. Amyloid deposition as the central event in the aetiology of Alzheimer's disease. Trends in Pharmacological Sciences. 1991;**12**(10):383-388. DOI: 10.1016/0165-6147(91)90609-v

[125] Hardy JA, Higgins GA. Alzheimer's disease: The amyloid cascade hypothesis. Science. 1992;**256**(5054): 184-185. DOI: 10.1126/science.1566067

[126] Lannfelt L, Relkin NR, Siemers ER. Amyloid-ss-directed immunotherapy for Alzheimer's disease. Journal of Internal Medicine. 2014;**275**(3):284-295. DOI: 10.1111/joim.12168

[127] Small SA, Duff K. Linking Abeta and tau in late-onset Alzheimer's disease: A dual pathway hypothesis. Neuron. 2008;**60**(4):534-542. DOI: 10.1016/j.neuron.2008.11.007

[128] Leng F, Edison P. Neuroinflammation and microglial activation in Alzheimer disease: Where do we go from here? Nature Reviews. Neurology. 2021;**17**(3):157-172. DOI: 10.1038/s41582-020-00435-y

[129] Grimaldi LM, Zappala G, Iemolo F, Castellano AE, Ruggieri S, Bruno G, et al. A pilot study on the use of interferon beta-1a in early Alzheimer's disease subjects. Journal of Neuroinflammation. 2014;**11**:30. DOI: 10.1186/1742-2094-11-30

[130] Zheng C, Zhou XW, Wang JZ. The dual roles of cytokines in Alzheimer's disease: Update on interleukins, TNF-alpha, TGF-beta and IFN-gamma. Translational Neurodegeneration. 2016;**5**:7. DOI: 10.1186/s40035-016-0054-4

[131] Wang S, Mustafa M, Yuede CM, Salazar SV, Kong P, Long H, et al. Anti-human TREM2 induces microglia proliferation and reduces pathology in an Alzheimer's disease model. Journal of Experimental Medicine. 2020;**217**(9): e20200785. DOI: 10.1084/jem.20200785

[132] Li C, Zhao B, Lin C, Gong Z, An X. TREM2 inhibits inflammatory responses in mouse microglia by suppressing the PI3K/NF-kappaB signaling. Cell Biology International. 2019;**43**(4):360-372. DOI: 10.1002/cbin.10975

[133] Schlepckow K, Monroe KM, Kleinberger G, Cantuti-Castelvetri L, Parhizkar S, Xia D, et al. Enhancing protective microglial activities with a dual function TREM2 antibody to the stalk region. EMBO Molecular Medicine. 2020;**12**(4):e11227. DOI: 10.15252/emmm.201911227

[134] Wunderlich P, Glebov K, Kemmerling N, Tien NT, Neumann H, Walter J. Sequential proteolytic processing of the triggering receptor expressed on myeloid cells-2 (TREM2) protein by ectodomain shedding and gamma-secretase-dependent intramembranous cleavage. The Journal of Biological Chemistry. 2013;**288**(46): 33027-33036. DOI: 10.1074/jbc.M113.517540

[135] Kleinberger G, Yamanishi Y, Suarez-Calvet M, Czirr E, Lohmann E, Cuyvers E, et al. TREM2 mutations implicated in neurodegeneration impair cell surface transport and phagocytosis. Science Translational Medicine. 2014;**6**(243):243ra86. DOI: 10.1126/scitranslmed.3009093

[136] Xiang X, Werner G, Bohrmann B, Liesz A, Mazaheri F, Capell A, et al. TREM2 deficiency reduces the efficacy of immunotherapeutic amyloid clearance. EMBO Molecular Medicine. 2016;**8**(9):992-1004. DOI: 10.15252/emmm.201606370

[137] Howard R, Liu KY. Questions EMERGE as biogen claims aducanumab turnaround. Nature Reviews. Neurology. 2020;**16**(2):63-64. DOI: 10.1038/s41582-019-0295-9

[138] Kurosawa Y, Degrauw TJ, Lindquist DM, Blanco VM, Pyne-Geithman GJ, Daikoku T, et al. Cyclocreatine treatment improves cognition in mice with creatine transporter deficiency. The Journal of Clinical Investigation. 2012;**122**(8):2837-2846. DOI: 10.1172/JCI59373

[139] Da Mesquita S, Louveau A, Vaccari A, Smirnov I, Cornelison RC, Kingsmore KM, et al. Functional aspects of meningeal lymphatics in ageing and Alzheimer's disease. Nature. 2018;**560**(7717):185-191. DOI: 10.1038/s41586-018-0368-8

[140] Da Mesquita S, Papadopoulos Z, Dykstra T, Brase L, Farias FG, Wall M, et al. Meningeal lymphatics affect microglia responses and anti-Abeta immunotherapy. Nature. 2021;**593**(7858):255-260. DOI: 10.1038/s41586-021-03489-0

[141] Mazanetz MP, Fischer PM. Untangling tau hyperphosphorylation in drug design for neurodegenerative diseases. Nature Reviews. Drug Discovery. 2007;**6**(6):464-479. DOI: 10.1038/nrd2111

[142] Wang JZ, Gong CX, Zaidi T, Grundke-Iqbal I, Iqbal K. Dephosphorylation of Alzheimer paired helical filaments by protein phosphatase-2A and -2B. The Journal of Biological Chemistry. 1995;**270**(9):4854-4860. DOI: 10.1074/jbc.270.9.4854

[143] Zheng J, Tian N, Liu F, Zhang Y, Su J, Gao Y, et al. A novel dephosphorylation targeting chimera selectively promoting tau removal in tauopathies. Signal Transduction and Targeted Therapy. 2021;**6**(1):269. DOI: 10.1038/s41392-021-00669-2

Chapter 7

The circRNA and Role in Alzheimer's Disease: From Regulation to Therapeutic and Diagnostic Targets

Wen Li and Guohua Jin

Abstract

Alzheimer's disease (AD) is a devastating neurodegenerative disorder and the most common form of dementia worldwide. Although the great progress on the prevention and treatment of AD, no effective therapies are available as yet. With the increasing incidence of AD, it has brought a growing burden to the family and society. Histopathologically, AD is characterized by the presence of myloid β (Aβ) plaques composed of Aβ and neurofibrillary tangles (NFTs) composed of hyper-phosphorylated tau proteins, which lead to neuronal loss. However, the full spectrum of precise molecular mechanism that contribute to AD pathogenesis remains largely unknown. circular RNAs (circRNAs) are a novel class of endogenous non-coding RNAs that play a vital role in post-transcriptional regulation. Recent reports showed circRNAs to be an important player in the development of neurodegenerative diseases like AD. In this chapter, we review recent progress on understanding the role of circRNAs in AD, and many studies implicating specific circRNAs in the development of the disease. Moreover, we explore the potential promise of these findings for future diagnosis and treatment.

Keywords: Alzheimer's disease, circular RNA, molecular mechanism, therapy

1. Introduction

Non-coding RNAs (ncRNAs) are a broad spectrum of functional RNA molecules that are transcribed from DNA but not translated into proteins [1]. The discovery of microRNAs (miRNAs) in 1993 followed by developments and discoveries in small RNA biology have hinted the importance of RNA in the post-transcriptional regulation of genes, especially in eukaryotes [2, 3]. With the development of high-throughput RNA sequencing technologies and bioinformatics methods, thousands of new ncRNAs have been discovered. Circular RNAs (circular RNAs, circRNAs) are a novel class of highly conservative endogenous ncRNA generated by pre-mRNA back splicing, which is characterized by a covalently closed-loop structure (**Figure 1**) [4]. It was originally discovered that circRNAs are universally present in human and mouse, but then they were found to be common across essentially all eukaryotes [5]. Although most circRNAs are generally expressed at low levels, some of them are more abundant than their linear counterparts and often exhibit

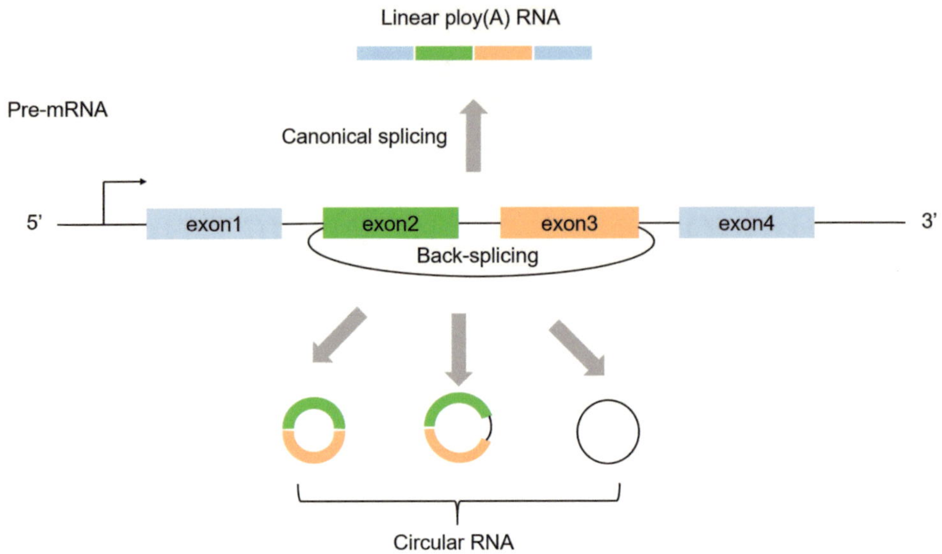

Figure 1.
Back-splicing and canonical splicing of a single pre-mRNA.

cell-type-specific and tissue-specific patterns [6, 7]. CircRNA levels are highly enriched in the brain, both embryonic and adult stage, and many of them play important roles in synaptic plasticity and neuronal function [8, 9]. Accumulating data suggest key roles for circRNAs in Alzheimer's disease (AD) by affecting mechanisms such as neuronal plasticity, autophagy, apoptosis, and inflammation [10]. This review covers the expression and function of circRNAs in the brain, as well as their role involvement in initiation and progression of AD.

2. Characteristics of circRNAs

2.1 Abundance of circRNAs

In the 1970s, circRNAs were first discovered in eukaryotic cytoplasm, but due to their low abundance and circular isoforms, these RNAs were perceived as mis-splicing occurrences [11]. With the advancement of high-throughput sequencing technology, circRNAs have been found to be abundant and widespread not only in metazoans including mice, Drosophila, and zebrafish but also in protists, fungi, and plants [7]. Although most of these covalently linked transcripts generally are expressed at low levels, in some cases, their abundance can exceed that of related linear mRNAs due to higher expression accumulation [12, 13]. For example, the expression of CDR1as within the brain is highly and independent of the expression of its linear isoform [14]. You et al. [15] discovered that the expression of some circRNAs exhibited strong up-regulation in brain during development, and their expression independent of their host linear transcripts. Jeck et al. [16] found that the abundance of some circRNAs exceeded associated linear mRNA by >10-fold in human fibroblasts.

2.2 Stability of circRNAs

CircRNAs are found mostly in the cytoplasm, and most undesired splicing products accumulate at the transcription site [17]. Because they do not have 5′-3′ polarities and polyadenylated tails, which make them much more stable than linear RNA and resistant to RNase R, an exonuclease that efficiently degrades linear RNAs [13]. The average half-life of circRNAs in cells exceeds 48 hours, while mRNAs only maintaine for the average of 10 hours [14]. Besides, circRNAs may also be sensitive to many other RNases, such as RNase A, RNase T1, and RNase T2 [18], which suggests that circRNA may serve as an ideal biomarker for a variety of disease. Recent studies have shown that circRNAs are enriched and stable in exosomes, which can be transported to distant tissues and organs via exosomes [19]. Moreover, circRNAs can be detected in blood cells like red blood cells, white blood cells, and platelets [20]. Besides blood, circRNAs can also be detected in other bodily fluids, such as saliva and seminal plasma [14, 21].

2.3 Profile and localization of circRNA

The expression of circRNA has tissue specificity, subcellular location specificity and developmental stage specificity. It is reported that circRNAs in *Drosophila*, mice, and humans are highly enriched in the nervous system [22]. RNA sequencing of human adult and fetal tissues showed that up to 50% of circRNAs were tissue-specific and development-specific fashion, and the number and expression levels of circRNAs were higher in fetal tissue than adult tissue [23]. Notably, the expression of circRNAs within the brain is highly specific and increases during neuronal differentiation and development, which may be involved in brain diseases [24]. CircRNAs are mostly found in the cytoplasm, and many of them in neurons localized to axons, dendrites, and synaptosomes, which is modulated by neuronal activity [15]. Reports show that compared with total brain RNA, circRNAs in murine synaptoneurosome fractions and micro-dissected neuropil from hippocampal slices were more enriched in cytoplasmic RNA [25]. This is supported by the detection of circRNAs in cultured hippocampal neurons and hippocampal slices [15]. Moreover, some circRNAs show a regulated switch in their nuclear and cytoplasmic positioning during development [24].

2.4 Classification of circRNAs

According to different combinations of sequences and domains, circRNAs can be divided into three categories: exonic circRNAs (EciRNAs), intronic circRNAs (CiRNAs) and exon-intron circRNAs (EIciRNA) (**Figure 2**) [26]. There are three hypothetical models explaining the formation of exonic circRNAs. Most circRNAs are formed by exon skipping during pre-mRNA transcription to produce specific regions, called lariat structures. Lariat structures contain exons, in which the intron sequence is then removed by splicing [27]. Other than exon skipping, due to the presence of reverse complement sequences in introns of pre-mRNA, circular structures can be formed by base-pairing between two introns, and some introns are then removed [28].

During the biogenesis of circRNAs, some RNA binding proteins (RBPs) are considered to participate in the circularization of circRNA, such as Quaking, Muscleblind and Fused-in sarcoma [29]. In some cases, during the formation of EciRNAs, introns that surround the exons are not removed, and EIciRNAs are generated [30]. The formation of CiRNAs depends on a consensus motif containing a conserved 7-nucleotide GU-rich motif at the 5′ splicing site and the 11-nucleotide C-rich motif at the 3′-branch site. it bypasses the action of the debranching enzyme, then generates a linear intron and form a circularized RNA lariat, leading to the production of ciRNA [9, 31]. EciRNAs mainly locate in the cytoplasm, which is the focus of current research, accounting for almost 80% of the total circRNAs [6]. CiRNAs and EIciRNAs locate in the nucleus, and regulate the expression of their parental genes [32].

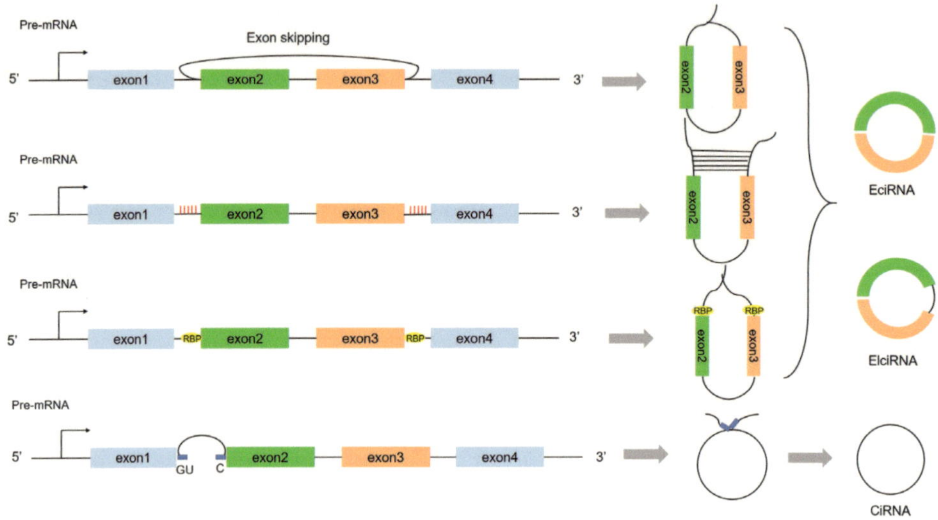

Figure 2.
Formation and classification of circRNAs.

3. Biological functions of circRNAs

circRNAs play key role in gene regulation at the post-transcriptional or transcription level, thereby affecting the level of gene expression. Here, we will introduce how circRNAs work at the molecular level and the underlying mechanisms involved in the interaction with other molecules (**Figure 3**).

3.1 circRNAs can act as microRNA (miRNA) sponges

Some long non-coding RNAs were described as a sponge for miRNAs, which can regulate the level or activity of miRNAs by selective sponging [33]. It was

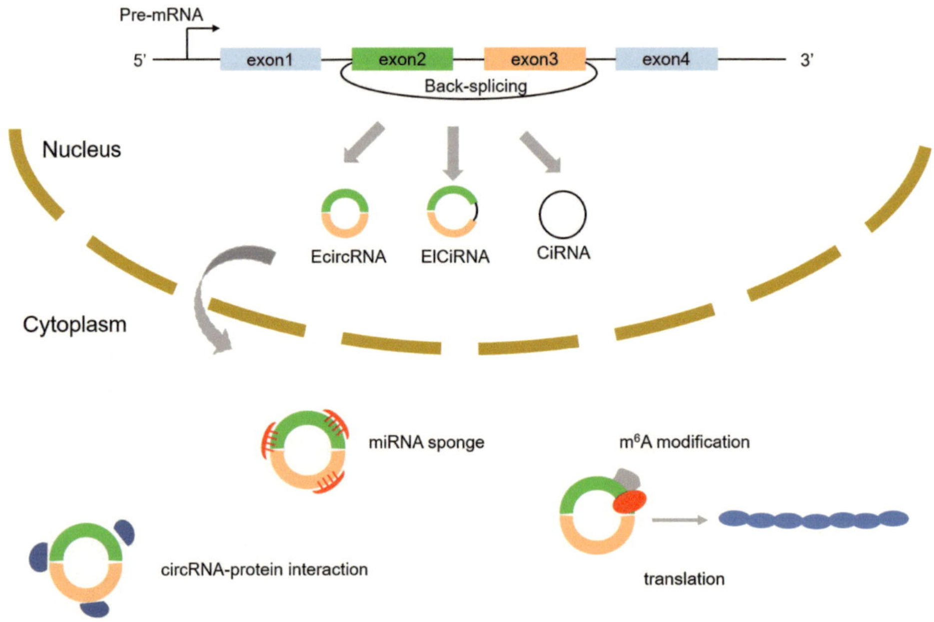

Figure 3.
Functions of circRNAs.

initially observed that some circRNAs have many miRNA-binding sites, leading to speculation that these molecules could act as miRNA sponges. Some circRNAs possess multiple binding sites for specific miRNAs, and some circRNAs harbor many different types of miRNA binding sites. For example, circRNA Cdr1as (ciRS-7) is extensively expressed in the mammalian brain and upregulated during neuronal development, which harbors 74 seed binding sites for miR-7 [34]. Intriguingly, miR-7 has been implicated, as a key regulator to modulate the expression of several oncogenes [35] and hold potential for slowing Parkinson's disease (PD) progression [36]. Similarly, circ-SRY (sex-determining region Y), is a master regulator of mammalian sex determination and specifically expressed in testis, which has 16 binding sites for miR-138 [37, 38]. Additionally, circHIPK3 is observed to sponge to 9 miRNAs with 18 potential binding sites [39]. cir-ITCH may act as a sponge of miR-7, miR-17, and miR-214 [40]. circHIPK3 was reported to bind to miR-124, miR-30a, and miR-558 [39, 41, 42]. So far, circRNA-miRNA axis in diseases have been expanded [43], and can be used as an advanced molecular technology to simulate or manufacture therapeutic agents, which indicates that this regulatory function of circRNAs should be a hotspot in the field of RNA. For instance, circHomer1 is upregulated in Hepatocellular carcinoma and regulates cell proliferation, migration, and invasion by inhibition of miR-1322 [44]. circTLK1 is upregulated during the acute period after focal ischemia, which can be functioned as an endogenous miR-335-3p sponge, leading to neuronal injury and neurological deficits [45]. Through deep RNA sequencing, novel_circ_0003012 and mmu-miR-298-3p were identified dysregulated in the hippocampus of APP/PS1 mice. Besides, novel_circ_0003012/mmu-miR-298-3p axis may regulate the pathological mechanism of AD by the cGPM-PKG signaling pathway [46].

3.2 circRNAs can interact with proteins

The most well-known proteins interacting with RNA molecules are the RBPs. RBPs are a large class of over 2000 proteins, that interact with transcripts to participate in forming ribonucleoprotein (RNP) complexes to influence the RNA fate [47]. Many circRNAs are predicted to interact with RBPs, although bioinformatic analyses of circRNA sequences revealed very little enrichment in binding sites of RBPs [48]. Human antigen R (HuR), an extensively studied RBP, regulates protein expression patterns by associating with a wide range of noncoding RNAs (ncRNAs), including miRNAs, long ncRNAs (lncRNAs), and circRNAs [49]. Li et al. [50] found that circPABPN1 blocked HuR binding to Atg16l1 mRNA, and represses HuR-induced ATG16L1 translation, thereby modulating Autophagy in the Intestinal Epithelium. Chen et al. [51] showed that oncogenic circAGO2 physically interacts with HuR, resulting in repression of AGO2/miRNA-mediated gene silencing during cancer progression. Quaking (QKI) is a member of the STAR family of KH domain-containing RNA-binding proteins, which is involved in pre-mRNA splicing, microRNA regulation, and formation of circRNA [52]. Gupta et al. [53] found that overexpression of Quaking 5 (Qki5) strongly attenuates doxorubicin-induced apoptosis and atrophy in cardiomyocytes via regulating a set of cardiac circRNAs. Zhu et al. [54] discovered that Qki5 is significantly downregulated in Hepatocellular carcinoma tissues, leading to the reduction of circZKSCAN1. Furthermore, circ-Foxo3 was observed to function as a scaffold to regulate the expression of its binding proteins by modulating protein–protein interaction. For example, circ-Foxo3 interacts with p21 and CDK2, promoting the inhibition of CDK2 by p21, and regulating cell cycle progression [17]. circ-Foxo3 can bind to p53 and Mdm2, to promote Mdm2-induced p53 ubiquitination and subsequent degradation, resulting in increased levels of Foxo3 protein [48]. circ-Foxo3 can interact with

ID-1, E2F1, FAK, and HIF1α, leading to these proteins retaining in the cytoplasm and no longer exerting their anti-senescent and anti-stress roles [55].

3.3 m^{6}A modification regulates circRNA translation

N6-methyladenosine (m^{6}A), the most prevalent internal RNA modification in mammalian cells, regulates RNA transcription, processing, splicing, degradation, and translation [56–58]. m^{6}A modification occurs by RNA methylation on the sixth N atom of adenylate (A) in RNAs [59]. m^{6}A modification sites tend to be found in the stop codon and 3′ untranslated region with a consensus sequence RRACH (in which R represents A or G and H represents A, C or U) [60]. The regulation function of m^{6}A is consisted of three factors referred to as "writers", "erasers" and "readers" [61]. m^{6}A "writers" are proteins involved in the formation of the methyltransferase complex, including methyltransferase-like 3/14/16 proteins (METTL3/14/16), Wilms tumor 1-associated protein (WTAP), RNA-binding motif protein 15/15B (RBM15/15B), and Vir-like m^{6}A methyltransferase associated (VIRMA, also known as KIAA1429) [62, 63].

m^{6}A methylation is dynamic and, and can be reversed by some demethylases (erasers). Erasers include FTO and AlkB homolog 3/5 (ALKBH3/5) [63, 64]. m^{6}A regulates gene expression through m^{6}A recognition factors, known as "readers," including YT521-B homology YTH domain family (YTHDF1/2/3), YTH domain containing 1 (YTHDC1/2), heterogeneous nuclear ribonucleoproteins (HNRNPs), eukaryotic translation initiation factor 3 (eIF3), and insulin-like growth factor-2 mRNA-binding proteins 1/2/3 (IGF2BP1/2/3) [61, 65].Recent studies have identified that m6A-modified circRNAs are related with pathophysiological processes. For example, m6A-modified RNA immunoprecipitation sequencing (m6A-RIP-seq) and RNA sequencing (RNA-seq) revealed the level of m6A abundance in total circRNAs was decreased in the lens epithelium cells (LECs) from cortical type of ARCs (ARCCs), and ALKBH5 was significantly upregulated [66]. Sun et al. [67] found that.

m6A modification are present on circPVRL3, which promoted gastric cancer cell proliferation. Huang et al. [68] found that circSTAG1 can bind ALKBH5 to inhibit its nuclear entry and increase the level of m6A modification of RNA, which attenuated depressive-like behaviors.

4. Role of circRNAs in AD

There is rising recognition that ncRNAs differences in the context of AD have yielded insight into the pathogenic mechanisms underlying this disease as well as biomarkers and potential therapeutic targets. Here, we provide the latest information on potential circRNAs involved in AD pathology.

4.1 AD pathogenesis

Reported histopathological characteristics of AD are Aβ plaques and NFTs, composed of Aβ protein accumulation and phosphorylated tau protein (p-tau) [69].

Amyloid pathogenesis starts with altered cleavage of amyloid precursor protein (APP) by β-secretases (BACE1) and γ-secretases, leading to the production of Aβ, which is then dumped into the extracellular space [70]. Consequently, accumulating Aβ forms Aβ oligomers and gradually polymerizes into amyloid fibrils that aggregate into plaques [71]. Tau is a microtubule-associated protein in neurons and plays an important role in maintaining the stability of microtubules [72]. Abnormal phosphorylation of tau makes it insoluble, reduces its ability to bind tubulin and promote microtubule assembly, and makes it self-associate into paired helical filament [73]. Additionally, microgliosis is consistently found around plaques in the brain [74]. This facilitates microglial activation and inflammatory response, and contributes to neuritic damage.

4.2 Aβ production and clearance in AD

Except for some antisense transcripts and microRNAs involved in accumulation, oligomerization, aggregation and formation of Aβ plaques, recent studies have shown that circRNAs may play a role in the of production and clearance Aβ [75, 76]. For example, circHDAC9 acted as a miR-138 sponge, decreasing miR-138 expression, Inhibiting the production of Aβ, and alleviating synaptic and learning/memory deficits in APP/PS1 mice. Moreover, circHDAC9 was remarkably decreased in the serum of both mild cognitive impairment and AD patients [77]. Shi et al. [78] found that ciRS-7 promotes the expression of UCHL1, reduces the protein levels of APP and BACE1 by promoting their degradation, and inhibits translation of NF-κB, thereby reducing the generation of Aβ. Shi et al. [79] demonstrated that circAβ-a, containing the corresponding Aβ coding sequence, served as a template for the synthesis of a novel Aβ-containing Aβ175 polypeptide in both cultured cells and human brain. Utilizing deep RNA sequencing, Zhang et al. [80] observed that there are 235 significantly dysregulated circRNA transcripts in a 7-month-old senescence-accelerated mouse prone 8. Additionally, circRNA-related ceRNA networks in this AD mouse model were mainly involved in the regulation of Aβ clearance.

4.3 Neuroinflammation in AD

A number of studies have proven that in addition to Aβ and NFTs, neuroinflammation is exhibited in the brains of AD patients and contributes to the pathogenesis of this disease [81, 82]. Not only Aβ can activate the microglia, but also Tau protein can trigger inflammation through interaction with microglia. Due to the accumulation of Aβ and hyperphosphorylation of Tau, microglia are persistently activated, which produce inflammatory cytokines and chemokines, contributing to the neuroinflammation process [81, 83]. For instance, TNF-α can stimulate γ-secretase activity, which results in increased levels of Aβ and the following cognitive decline in AD [84]. IL-1 increases generation of Aβ and phosphorylation of tau protein, leading to dysfunction of the cholinergic system [85]. Studies have shown that CCL2 and CCL5 expression are increased in the AD brain. Up to now, there are few reports in this field. Wang et al. [86] found that in OGD-activated microglia, circPTK2 regulates neuronal apoptosis via sponging miR-29b. it can be inferred that circRNAs may be involved in the activation of AD microglia. One study on circRNAs involved in neuroinflammation indicates that circ_0000950 inhibits miR-103 expression and increases prostaglandin-endoperoxide synthase 2 (PTGS2) expression in AD models. Moreover, circ_0000950 promotes neuron apoptosis, neurite outgrowth, and affects the level of IL-1β, IL-6 and TNF-α via directly sponging miR-103 in AD [87].

4.4 Oxidative stress in AD

Emerging evidences demonstrate that oxidative stress has been recognized as a contributing factor in the progression of AD. It has been confirmed that elevated levels of Aβ are associated with increased levels of oxidation products, and protein and lipid oxidation was observed in brain regions rich in Aβ [88, 89]. Moreover, due to the role of protein Tau both in the modulation microtubule dynamics and morphology and physiology of neurons, Tau alteration would constitute to a target for oxidative stress in AD [90, 91]. Recently, important regulatory roles of some circRNAs in the oxidative stress have been identified. Previous studies have shown that panax notoginseng saponins (PNS) could protect neurons in AD brain from oxidative stress damage injury via attenuating the production of 8-hydroxydeoxy-guanosine (8-OHdG), enhanceing the expressions and activities of superoxide dismutase (SOD), catalase (CAT), and glutathione peroxidase (GSH-PX) [92]. Through circRNA Microarray, it was found that PNS treatment leads to five circRNAs upregulation and two circRNAs downregulation. Next, mmu_circRNA_013636 and mmu_circRNA_012180 were selected, and GO and KEGG analyses were showed that mmu_circRNA_013636 and mmu_circRNA_012180 were involved in AD-associated biological process [93]. Based on these results, the mmu_circRNA_013636 and mmu_circRNA_012180 may be associated with the mechanisms by which PNS attenuates AD progression, and may be highly related to the regulation of oxidative stress. Zhu et al. [94] found that the expression level of circular ribonucleic acid 0001588 was suppressed in model of AD, which promoted cell growth, reduced levels of lactate dehydrogenase, caspase-3, and caspase-9. Besides, circular ribonucleic acid 0001588 reduced reactive oxygen species production via activation of the silent information regulator 1 pathway.

4.5 Autophagy in AD

Substantial studies reveal that deficits in autophagy are involved in AD pathogenesis. Defective autophagy and mitophagy, which is responsible for synaptic dysfunction and cognitive deficits, are triggered by Aβ and Tau accumulation [95]. Recently, several reports have described potential roles for circRNAs in autophagosome assembly or vesicular transport-mediated pathways [96]. For example, Chen et al. [97] found that circNF1–419 regulates autophagy through PI3K-I/Akt-AMPK-mTOR and PI3K-I/Akt–mTOR signaling pathways, and reduces the expression of AD marker proteins Tau, p-Tau, Aβ1–42, and APOE in AD-like mice. Using circRNA microarray, GO analysis revealed that mmu_circRNA_017963 is highly associated with autophagosome assembly, exocytosis, apoptotic process, transport and RNA splicing in an AD mouse model. Moreover, KEGG pathway analysis indicated that mmu_circRNA_017963 was strongly related with synaptic vesicle cycle, spliceosome, glycosaminoglycan, and SNARE interactions in vesicular transport [98]. All of these biological processes are reported to play an important role in the development of AD [98, 99].

4.6 Therapeutic targets and diagnostic biomarkers in AD

Due to the high stability with covalently closed continuous loop, circRNAs are not sensitive to ribonucleases, such as RNase R, and have a longer half-life compared to linear RNAs [100]. An accumulating number of studies have shown that dysregulated circRNAs are significantly related to AD, which are considered to be potential biomarkers. For instance, Dube et al. [101] found that circRNA expression significantly associated with the diagnosis of AD, the severity of clinical dementia, and the severity of neuropathology. Lo et al. [102] profiled circRNA expression at different AD stages in brain samples from four brain regions: anterior prefrontal cortex, superior temporal lobe, parahippocampal gyrus and inferior frontal gyrus using a public RNA-sequencing dataset. There are 147 differentially expressed circRNAs to be found in the four regions, and most circRNAs in AD patients with severe symptoms are enriched in the parahippocampal gyrus. This finding could help to distinguish the disease severity of patients, and further implying that circRNAs may serve as biomarkers of AD. In addition, circRNAs can stably exist in blood plasma and cerebrospinal fluid. Liu et al. [103] discovered that hsa_circ_0003391 is significantly downregulated in the peripheral blood, and closely related to clinical features of patients with AD. By microarray, Li et al. [104] found 112 circRNAs were upregulated and 51 circRNAs were downregulated in cerebrospinal fluid of AD patients. Among the up-regulated circRNAs, circ-AXL was negatively correlated with Aβ42 and positively correlated with t-Tau and p-tau, suggesting it hold the clinical value for predicting disease risk and disease severity of AD. Moreover, research on circRNAs sponges may help to design and develop effective artificial sponges to regulate disease progression. As a stable and effective miRNA inhibitor, artificial miRNA sponge technology may be a new strategy for RNA gene therapy in the future.

5. Conclusions and future perspectives

circRNAs have gained increased attention because of their involvement in different biological processes. With the rapid progress of high throughput sequencing and bioinformatics technology, multiple circRNAs have demonstrated to be closely associated with various diseases. Although the function and modulation of circRNA has not been clearly understood, studies have started to excavate effect of AD-related circRNAs, which brought us many surprising findings. As a class of stable RNA, circRNAs have natural advantages and may play vital roles as therapeutic targets and prognostic factors for AD. Recently, with the emergence of CRISPR-Cas13d screening tools [105], lipid nanoparticle (LNP) delivery system [106], the in vitro engineered preparation of circRNA can be realized, and make the application of circRNA in clinical therapy possible. In addition, Lavenniah et al. [107] constructed a circmiR sponge targeting the known cardiac pro-hypertrophic miRs-132 and miRs-212, and delivered it to cardiomyocytes *in vivo by* Adeno-associated viruses (AAVs). Subsequently, the hypertrophic characteristics of the disease were attenuated, thus supporting the therapeutic potential of Engineered circRNAs. However, there are some some questions that deserve attention. Firstly, most of the current studies on circRNAs rely on the results of RNA-sequencing and microarray. There are significant differences between the output of different algorithms, emphasizing that these circRNAs need urther validation. Secondly, many circRNAs have been identified to be differentially expressed at different developmental stages, but the precise mechanisms are still not clear. Thirdly, at present, most studies on circRNAs mainly focus on miRNA sponge, and there are few studies on other mechanisms. In conclusion, circRNA research is still in its infancy and their molecular mechanism and functional role need to be further elucidated. circRNAs are widely involved in the regulation of physiological and pathophysiological processes, and may have the potential to be new biomarkers and novel therapeutic targets.

Author details

Wen Li[1,2,3*] and Guohua Jin[1,2,3]

1 Department of Human Anatomy, Institute of Neurobiology, Medical School of Nantong University, Nantong, Jiangsu, P.R. China

2 Key Laboratory of Neuroregeneration of Jiangsu and Ministry of Education, Nantong, Jiangsu, P.R. China

3 Co-Innovation Center of Neuroregeneration, Medical School of Nantong University, Nantong, Jiangsu, P.R. China

*Address all correspondence to: 1554371938@qq.com

References

[1] Poller W, et al. Non-coding RNAs in cardiovascular diseases: diagnostic and therapeutic perspectives. Eur Heart J, 2018, 39(29): 2704-2716.

[2] Lee RC, Feinbaum RL, and Ambros V. The *C. elegans* heterochronic gene lin-4 encodes small RNAs with antisense complementarity to lin-14. Cell, 1993, 75(5): 843-854.

[3] Yang JD, Liu MC, and Kisiel JB. Circulating Tumor DNA and Hepatocellular Carcinoma. Semin Liver Dis, 2019, 39(4): 452-462.

[4] Wu X, et al. Circular RNA: A novel potential biomarker for skin diseases. Pharmacol Res, 2020, 158(104841.

[5] Salzman J. Circular RNA Expression: Its Potential Regulation and Function. Trends Genet, 2016, 32(5): 309-316.

[6] Zhang P, et al. Circular RNA Regulation of Myogenesis. Cells, 2019, 8(8):

[7] Li X, Yang L, and Chen LL. The Biogenesis, Functions, and Challenges of Circular RNAs. Mol Cell, 2018, 71(3): 428-442.

[8] Hanan M, Soreq H, and Kadener S. CircRNAs in the brain. RNA Biol, 2017, 14(8): 1028-1034.

[9] Mehta SL, Dempsey RJ, and Vemuganti R. Role of circular RNAs in brain development and CNS diseases. Prog Neurobiol, 2020, 186(101746.

[10] Zhang Y, et al. Exploring the regulatory roles of circular RNAs in Alzheimer's disease. Transl Neurodegener, 2020, 9(1): 35.

[11] Sanger HL, et al. Viroids are single-stranded covalently closed circular RNA molecules existing as highly base-paired rod-like structures. Proc Natl Acad Sci U S A, 1976, 73(11): 3852-3856.

[12] Greene J, et al. Circular RNAs: Biogenesis, Function and Role in Human Diseases. Front Mol Biosci, 2017, 4(38.

[13] Xiao MS, Ai Y, and Wilusz JE. Biogenesis and Functions of Circular RNAs Come into Focus. Trends Cell Biol, 2020, 30(3): 226-240.

[14] Li HM, Ma XL, and Li HG. Intriguing circles: Conflicts and controversies in circular RNA research. Wiley Interdiscip Rev RNA, 2019, 10(5): e1538.

[15] You X, et al. Neural circular RNAs are derived from synaptic genes and regulated by development and plasticity. Nat Neurosci, 2015, 18(4): 603-610.

[16] Jeck WR, et al. Circular RNAs are abundant, conserved, and associated with ALU repeats. RNA, 2013, 19(2): 141-157.

[17] Du WW, et al. Identifying and Characterizing circRNA-Protein Interaction. Theranostics, 2017, 7(17): 4183-4191.

[18] Vincent HA and Deutscher MP. Substrate recognition and catalysis by the exoribonuclease RNase R. J Biol Chem, 2006, 281(40): 29769-29775.

[19] Kim KM, et al. RNA in extracellular vesicles. Wiley Interdiscip Rev RNA, 2017, 8(4):

[20] Ouyang Q, et al. Microarray Expression Profile of Circular RNAs in Peripheral Blood Mononuclear Cells from Rheumatoid Arthritis Patients. Cell Physiol Biochem, 2017, 42(2): 651-659.

[21] Jafari Ghods F. Circular RNA in Saliva. Adv Exp Med Biol, 2018, 1087(131-139.

[22] Zheng S, et al. CircRNA-Protein Interactions in Muscle Development and Diseases. Int J Mol Sci, 2021, 22(6):

[23] Xu T, et al. Circular RNA expression profiles and features in human tissues: a study using RNA-seq data. BMC Genomics, 2017, 18(Suppl 6): 680.

[24] Patop IL, Wust S, and Kadener S. Past, present, and future of circRNAs. EMBO J, 2019, 38(16): e100836.

[25] Rybak-Wolf A, et al. Circular RNAs in the Mammalian Brain Are Highly Abundant, Conserved, and Dynamically Expressed. Mol Cell, 2015, 58(5): 870-885.

[26] Ma S, et al. CircRNAs: biogenesis, functions, and role in drug-resistant Tumours. Mol Cancer, 2020, 19(1): 119.

[27] Ma Y, Liu Y, and Jiang Z. CircRNAs: A new perspective of biomarkers in the nervous system. Biomed Pharmacother, 2020, 128(110251.

[28] Fang Z, Jiang C, and Li S. The Potential Regulatory Roles of Circular RNAs in Tumor Immunology and Immunotherapy. Front Immunol, 2020, 11(617583.

[29] Tang X, et al. Review on circular RNAs and new insights into their roles in cancer. Comput Struct Biotechnol J, 2021, 19(910-928.

[30] Shang Q, et al. The novel roles of circRNAs in human cancer. Mol Cancer, 2019, 18(1): 6.

[31] Li J, et al. CircRNAs: a new target for the diagnosis and treatment of digestive system neoplasms. Cell Death Dis, 2021, 12(2): 205.

[32] Floris G, et al. Regulatory Role of Circular RNAs and Neurological Disorders. Mol Neurobiol, 2017, 54(7): 5156-5165.

[33] Han TS, et al. Epigenetic Associations between lncRNA/circRNA and miRNA in Hepatocellular Carcinoma. Cancers (Basel), 2020, 12(9):

[34] Guo Z, et al. Biogenesis, Features, Functions, and Disease Relationships of a Specific Circular RNA: CDR1as. Aging Dis, 2020, 11(4): 1009-1020.

[35] Hansen TB, Kjems J, and Damgaard CK. Circular RNA and miR-7 in cancer. Cancer Res, 2013, 73(18): 5609-5612.

[36] D'ambra E, Capauto D, and Morlando M. Exploring the Regulatory Role of Circular RNAs in Neurodegenerative Disorders. Int J Mol Sci, 2019, 20(21):

[37] Hansen TB, et al. Natural RNA circles function as efficient microRNA sponges. Nature, 2013, 495(7441): 384-388.

[38] Zhao ZJ and Shen J. Circular RNA participates in the carcinogenesis and the malignant behavior of cancer. RNA Biol, 2017, 14(5): 514-521.

[39] Zheng Q, et al. Circular RNA profiling reveals an abundant circHIPK3 that regulates cell growth by sponging multiple miRNAs. Nat Commun, 2016, 7(11215.

[40] Li F, et al. Circular RNA ITCH has inhibitory effect on ESCC by suppressing the Wnt/beta-catenin pathway. Oncotarget, 2015, 6(8): 6001-6013.

[41] Chen B, et al. Circular RNA circHIPK3 Promotes the Proliferation and Differentiation of Chicken Myoblast Cells by Sponging miR-30a-3p. Cells, 2019, 8(2):

[42] Li Y, et al. CircHIPK3 sponges miR-558 to suppress heparanase expression in bladder cancer cells. EMBO Rep, 2017, 18(9): 1646-1659.

[43] Su Q and Lv X. Revealing new landscape of cardiovascular disease through circular RNA-miRNA-mRNA axis. Genomics, 2020, 112(2): 1680-1685.

[44] Zhao M, et al. Circ-HOMER1 enhances the inhibition of miR-1322 on CXCL6 to regulate the growth and aggressiveness of hepatocellular carcinoma cells. J Cell Biochem, 2020, 121(11): 4440-4449.

[45] Wu F, et al. Circular RNA TLK1 Aggravates Neuronal Injury and Neurological Deficits after Ischemic Stroke via miR-335-3p/TIPARP. J Neurosci, 2019, 39(37): 7369-7393.

[46] Zhang Y, et al. CircRNA-ceRNA Network Revealing the Potential Regulatory Roles of CircRNA in Alzheimer's Disease Involved the cGMP-PKG Signal Pathway. Front Mol Neurosci, 2021, 14(665788.

[47] Schuschel K, et al. RNA-Binding Proteins in Acute Leukemias. Int J Mol Sci, 2020, 21(10):

[48] Huang A, et al. Circular RNA-protein interactions: functions, mechanisms, and identification. Theranostics, 2020, 10(8): 3503-3517.

[49] Abdelmohsen K, et al. Identification of HuR target circular RNAs uncovers suppression of PABPN1 translation by CircPABPN1. RNA Biol, 2017, 14(3): 361-369.

[50] Li XX, et al. Interaction between HuR and circPABPN1 Modulates Autophagy in the Intestinal Epithelium by Altering ATG16L1 Translation. Mol Cell Biol, 2020, 40(6):

[51] Chen Y, et al. Circular RNA circAGO2 drives cancer progression through facilitating HuR-repressed functions of AGO2-miRNA complexes. Cell Death Differ, 2019, 26(7): 1346-1364.

[52] Chen X, et al. The Emerging Roles of the RNA Binding Protein QKI in Cardiovascular Development and Function. Front Cell Dev Biol, 2021, 9(668659.

[53] Gupta SK, et al. Quaking Inhibits Doxorubicin-Mediated Cardiotoxicity Through Regulation of Cardiac Circular RNA Expression. Circ Res, 2018, 122(2): 246-254.

[54] Zhu YJ, et al. Circular RNAs negatively regulate cancer stem cells by physically binding FMRP against CCAR1 complex in hepatocellular carcinoma. Theranostics, 2019, 9(12): 3526-3540.

[55] Du WW, et al. Foxo3 circular RNA promotes cardiac senescence by modulating multiple factors associated with stress and senescence responses. Eur Heart J, 2017, 38(18): 1402-1412.

[56] Chen M and Wong CM. The emerging roles of N6-methyladenosine (m6A) deregulation in liver carcinogenesis. Mol Cancer, 2020, 19(1): 44.

[57] Cheng M, et al. The m(6)A methyltransferase METTL3 promotes bladder cancer progression via AFF4/NF-kappaB/MYC signaling network. Oncogene, 2019, 38(19): 3667-3680.

[58] Barbieri I, et al. Promoter-bound METTL3 maintains myeloid leukaemia by m(6)A-dependent translation control. Nature, 2017, 552(7683): 126-131.

[59] Zhou P, et al. Meclofenamic acid promotes cisplatin-induced acute kidney injury by inhibiting fat mass and obesity-associated protein-mediated m(6)A abrogation in RNA. J Biol Chem, 2019, 294(45): 16908-16917.

[60] Li Y, et al. The dynamics of FTO binding and demethylation from the m(6)A motifs. RNA Biol, 2019, 16(9): 1179-1189.

[61] Liu ZX, et al. Link Between m6A Modification and Cancers. Front Bioeng Biotechnol, 2018, 6(89.

[62] Zhang Y, et al. m6A modification in RNA: biogenesis, functions and roles in gliomas. J Exp Clin Cancer Res, 2020, 39(1): 192.

[63] Zhang L, et al. The role of N(6)-methyladenosine (m(6)A) modification in the regulation of circRNAs. Mol Cancer, 2020, 19(1): 105.

[64] Yi YC, et al. Novel insights into the interplay between m(6)A modification and noncoding RNAs in cancer. Mol Cancer, 2020, 19(1): 121.

[65] Zhang C, et al. Reduced m6A modification predicts malignant phenotypes and augmented Wnt/PI3K-Akt signaling in gastric cancer. Cancer Med, 2019, 8(10): 4766-4781.

[66] Li P, et al. Identification and Characterization of N6-Methyladenosine CircRNAs and Methyltransferases in the Lens Epithelium Cells From Age-Related Cataract. Invest Ophthalmol Vis Sci, 2020, 61(10): 13.

[67] Sun HD, et al. Down-regulation of circPVRL3 promotes the proliferation and migration of gastric cancer cells. Sci Rep, 2018, 8(1): 10111.

[68] Tang M and Lv Y. The Role of N(6)-Methyladenosine Modified Circular RNA in Pathophysiological Processes. Int J Biol Sci, 2021, 17(9): 2262-2277.

[69] Jiang L, et al. Exosomes in Pathogenesis, Diagnosis, and Treatment of Alzheimer's Disease. Med Sci Monit, 2019, 25(3329-3335.

[70] O'brien RJ and Wong PC. Amyloid precursor protein processing and Alzheimer's disease. Annu Rev Neurosci, 2011, 34(185-204.

[71] Chen GF, et al. Amyloid beta: structure, biology and structure-based therapeutic development. Acta Pharmacol Sin, 2017, 38(9): 1205-1235.

[72] Guo T, Noble W, and Hanger DP. Roles of tau protein in health and disease. Acta Neuropathol, 2017, 133(5): 665-704.

[73] Deture MA and Dickson DW. The neuropathological diagnosis of Alzheimer's disease. Mol Neurodegener, 2019, 14(1): 32.

[74] Hanslik KL and Ulland TK. The Role of Microglia and the Nlrp3 Inflammasome in Alzheimer's Disease. Front Neurol, 2020, 11(570711.

[75] Cieslik M, et al. Alterations of Transcription of Genes Coding Anti-oxidative and Mitochondria-Related Proteins in Amyloid beta Toxicity: Relevance to Alzheimer's Disease. Mol Neurobiol, 2020, 57(3): 1374-1388.

[76] Idda ML, et al. Noncoding RNAs in Alzheimer's disease. Wiley Interdiscip Rev RNA, 2018, 9(2):

[77] Lu Y, Tan L, and Wang X. Circular HDAC9/microRNA-138/Sirtuin-1 Pathway Mediates Synaptic and Amyloid Precursor Protein Processing Deficits in Alzheimer's Disease. Neurosci Bull, 2019, 35(5): 877-888.

[78] Shi Z, et al. The circular RNA ciRS-7 promotes APP and BACE1 degradation in an NF-kappaB-dependent manner. FEBS J, 2017, 284(7): 1096-1109.

[79] Mo D, et al. Circular RNA Encoded Amyloid Beta peptides-A Novel Putative Player in Alzheimer's Disease. Cells, 2020, 9(10):

[80] Zhang S, et al. Characterization of circRNA-Associated-ceRNA Networks in a Senescence-Accelerated Mouse Prone 8 Brain. Mol Ther, 2017, 25(9): 2053-2061.

[81] Sung PS, et al. Neuroinflammation and Neurogenesis in Alzheimer's Disease and Potential Therapeutic Approaches. Int J Mol Sci, 2020, 21(3):

[82] Sanchez-Sarasua S, et al. Can We Treat Neuroinflammation in Alzheimer's Disease? Int J Mol Sci, 2020, 21(22):

[83] Lian H, et al. Astrocyte-Microglia Cross Talk through Complement Activation Modulates Amyloid Pathology in Mouse Models of Alzheimer's Disease. J Neurosci, 2016, 36(2): 577-589.

[84] Hampel H, et al. A Path Toward Precision Medicine for Neuroinflammatory Mechanisms in Alzheimer's Disease. Front Immunol, 2020, 11(456.

[85] Su F, Bai F, and Zhang Z. Inflammatory Cytokines and Alzheimer's Disease: A Review from the Perspective of Genetic Polymorphisms. Neurosci Bull, 2016, 32(5): 469-480.

[86] Wang H, et al. Circular RNA circPTK2 regulates oxygen-glucose deprivation-activated microglia-induced hippocampal neuronal apoptosis via miR-29b-SOCS-1-JAK2/STAT3-IL-1beta signaling. Int J Biol Macromol, 2019, 129(488-496.

[87] Yang H, et al. Circular RNA circ_0000950 promotes neuron apoptosis, suppresses neurite outgrowth and elevates inflammatory cytokines levels via directly sponging miR-103 in Alzheimer's disease. Cell Cycle, 2019, 18(18): 2197-2214.

[88] Butterfield DA and Lauderback CM. Lipid peroxidation and protein oxidation in Alzheimer's disease brain: potential causes and consequences involving amyloid beta-peptide-associated free radical oxidative stress. Free Radic Biol Med, 2002, 32(11): 1050-1060.

[89] Butterfield DA. The 2013 SFRBM discovery award: selected discoveries from the butterfield laboratory of oxidative stress and its sequela in brain in cognitive disorders exemplified by Alzheimer disease and chemotherapy induced cognitive impairment. Free Radic Biol Med, 2014, 74(157-174.

[90] Tapia-Rojas C, et al. It's all about tau. Prog Neurobiol, 2019, 175(54-76.

[91] Cheignon C, et al. Oxidative stress and the amyloid beta peptide in Alzheimer's disease. Redox Biol, 2018, 14(450-464.

[92] Huang JL, et al. Neuroprotective Properties of Panax notoginseng Saponins via Preventing Oxidative Stress Injury in SAMP8 Mice. Evid Based Complement Alternat Med, 2017, 2017(8713561.

[93] Huang JL, et al. Identification of Differentially Expressed Profiles of Alzheimer's Disease Associated Circular RNAs in a Panax Notoginseng Saponins-Treated Alzheimer's Disease Mouse Model. Comput Struct Biotechnol J, 2018, 16(523-531.

[94] Zhu R, et al. The silent information regulator 1 pathway attenuates ROS-induced oxidative stress in Alzheimer's disease. J Integr Neurosci, 2020, 19(2): 321-332.

[95] Reddy PH and Oliver DM. Amyloid Beta and Phosphorylated Tau-Induced Defective Autophagy and Mitophagy in Alzheimer's Disease. Cells, 2019, 8(5):

[96] Zhang Y, et al. The Role of Non-coding RNAs in Alzheimer's Disease: From Regulated Mechanism to Therapeutic Targets and Diagnostic Biomarkers. Front Aging Neurosci, 2021, 13(654978.

[97] Diling C, et al. Circular RNA NF1-419 enhances autophagy to ameliorate senile dementia by binding Dynamin-1 and Adaptor protein 2 B1 in AD-like mice. Aging (Albany NY), 2019, 11(24): 12002-12031.

[98] Huang JL, et al. Comprehensive analysis of differentially expressed profiles of Alzheimer's disease associated circular RNAs in an Alzheimer's disease mouse model. Aging (Albany NY), 2018, 10(2): 253-265.

[99] Cui W, et al. Inhibition of PTEN Attenuates Endoplasmic Reticulum Stress and Apoptosis via Activation of PI3K/AKT Pathway in Alzheimer's Disease. Neurochem Res, 2017, 42(11): 3052-3060.

[100] Sheng R, et al. Circular RNAs and their emerging roles as diagnostic and prognostic biomarkers in ovarian cancer. Cancer Lett, 2020, 473(139-147.

[101] Dube U, et al. An atlas of cortical circular RNA expression in Alzheimer disease brains demonstrates clinical and pathological associations. Nat Neurosci, 2019, 22(11): 1903-1912.

[102] Lo I, et al. Linking the association between circRNAs and Alzheimer's disease progression by multi-tissue circular RNA characterization. RNA Biol, 2020, 17(12): 1789-1797.

[103] Liu L, et al. Identification of Circular RNA hsa_Circ_0003391 in Peripheral Blood Is Potentially Associated With Alzheimer's Disease. Front Aging Neurosci, 2020, 12(601965.

[104] Li Y, et al. Circular RNA expression profile of Alzheimer's disease and its clinical significance as biomarkers for the disease risk and progression. Int J Biochem Cell Biol, 2020, 123(105747.

[105] Li S, et al. Screening for functional circular RNAs using the CRISPR-Cas13 system. Nat Methods, 2021, 18(1): 51-59.

[106] Ickenstein LM and Garidel P. Lipid-based nanoparticle formulations for small molecules and RNA drugs. Expert Opin Drug Deliv, 2019, 16(11): 1205-1226.

[107] Lavenniah A, et al. Engineered Circular RNA Sponges Act as miRNA Inhibitors to Attenuate Pressure Overload-Induced Cardiac Hypertrophy. Mol Ther, 2020, 28(6): 1506-1517.

Chapter 8

Transcranial Red LED Therapy: A Promising Non-Invasive Treatment to Prevent Age-Related Hippocampal Memory Impairment

Claudia Jara, Débora Buendía, Alvaro Ardiles, Pablo Muñoz and Cheril Tapia-Rojas

Abstract

The hippocampus is an integral portion of the limbic system and executes a critical role in spatial and recognition learning, memory encoding, and memory consolidation. Hippocampal aging showed neurobiological alterations, including increased oxidative stress, altered intracellular signaling pathways, synaptic impairment, and organelle deterioration such as mitochondrial dysfunction. These alterations lead to hippocampal cognitive decline during aging. Therefore, the search for new non-invasive therapies focused on preserving or attenuating age-related hippocampal memory impairment could have of great impact on aging, considering the increasing life expectancy in the world. Red light Transcranial LED therapy (RL-TCLT) is a promising but little explored strategy, which involves red light LED irradiation without surgical procedures, safe and at a low cost. Nevertheless, the precise mechanism involved and its real impact on age-related cognitive impairment is unclear, due to differences in protocol, wavelength applied, and time. Therefore, in this chapter, we will discuss the evidence about RL-TCLT and its effects on the hippocampal structure and function, and how this therapy could be used as a promising treatment for memory loss during aging and in age-related diseases such as Alzheimer's Disease (AD). Finally, we will mention our advances in Red 630-light-Transcranial LED therapy on the hippocampus in aging and AD.

Keywords: aging, hippocampus, memory, LED therapy, mitochondria

1. Introduction

Aging is a biological process characterized by a general decline in cell function. Life expectancy is increasing and has turned aging into a social problem in the world. The brain is one of the organs that is most affected by age [1, 2], therefore new investigations into safe and non-invasive treatments to reduce age-related brain damage and subsequent cognitive impairment are of critical importance. The aging brain displays synaptic alterations that negatively affect cognitive capacity, especially memory. The hippocampus mediates the formation of new memories and age-related hippocampal dysfunction compromises learning and memory processes [3].

Interestingly, in hippocampal memory loss, mitochondrial dysfunction plays a central role. Synaptic and mitochondrial dysfunction are early events in aging, mutually influenced, triggering age-associated memory defects [4]. Then, the need arises to find new strategies that can help elderly people to pass a better old age, without forgetting their memories or their history.

A promising but little explored strategy is the application of non-invasive cell stimulation with specific light types. Photobiomodulation is the use of light to stimulate or regenerate organs and tissues. Red-near-infrared (800-1100 nm) and red (600 nm) wavelengths of light-emitting diodes (LED) have been used for a range of therapeutic purposes [5–7]. These wavelengths could penetrate through the skin and have the potential to improve the cellular function of compromised tissue [5, 7]. Red-near-infrared and red LED therapy involves the interaction of photons with molecules in the cells [5, 8, 9]. Specifically, Transcranial LED therapy (TCLT) defines the limited application of LED therapy to the brain. The LED light travels through the layers of the scalp and skull to reach brain cells [10–12]. The brain is commonly irradiated with red (RL) or near-infrared (NIR) light (600-1100 nm), with a total output power of 1-10,000 mW, a power density that has no thermal effects [9]. Several studies have reported the use of brain irradiation with red or near-infrared (600-1100 nm) LED improving tissue repair, blood flow, cicatrization, and recovery following trauma [12–14]; however, the results are variable due to differences in protocols and wavelengths, LED potential, stimulation time the tissue target, the animal model used, as well as the doses or treatment period [13, 15–17].

Diverse experimental and clinical studies have been performed to test transcranial LED therapy with promising results in brain function [9, 14, 16]. Thus, *in vivo* studies using 660 nm and 810 nm Red-light Transcranial LED therapy (RL-TCLT) in a mice model of aging induced by D-galactose in BALB/c mice improved spatial memory and increased mitochondrial function [18]. In transgenic AD mice, RL-LED treatment of the whole body recovered interstitial fluid flow, reduced Aβ deposition in the brain, and alleviate cognitive deficits [19]. Furthermore, studies in patients victims of severe traumatic brain injury (TBI) showed positive effects after RL-TCLT, enhancing their quality of life, by improving their memory, and decreasing affections such as pain, depression, nervousness, and insomnia (**Figure 1**) [12, 20, 21].

Also, complementary *in vitro* studies with 600–850 nm LED irradiation showed light absorption by the cytochrome c oxidase (COX) enzyme, the complex IV of the oxidative phosphorylation (OXPHOS) system located in the electron transport chain (ETC) from the mitochondria [5], leading to the upregulation of the mitochondrial respiratory capacity and increased ATP production [4, 22]. *In vitro* assays also propose that mitochondrial COX act as a photoreceptor that mediates the beneficial effects of photobiomodulation [23]. Nevertheless, until is unclear how COX mediates the beneficial effect regulating energy production, and for this reason, most of the reports concluded that the mechanism underlying the neuroprotective actions of RL-TCLT is not completely understood. More studies are required to determine the biological events that lead to neuroprotection or neuronal repair.

In this chapter, we will summarize the evidence about the studies using Red Light Transcranial LED therapy (RL-TCLT), mainly focused on their positive effect in the brain, and particularly in the hippocampal structure and function. In addition, we will discuss the possible mechanisms involved in the beneficial effects of RL-TCLT, putting particular emphasis on the mitochondria. Finally, we will briefly comment on our main finding using RL-TCLT, as a potential anti-aging therapy.

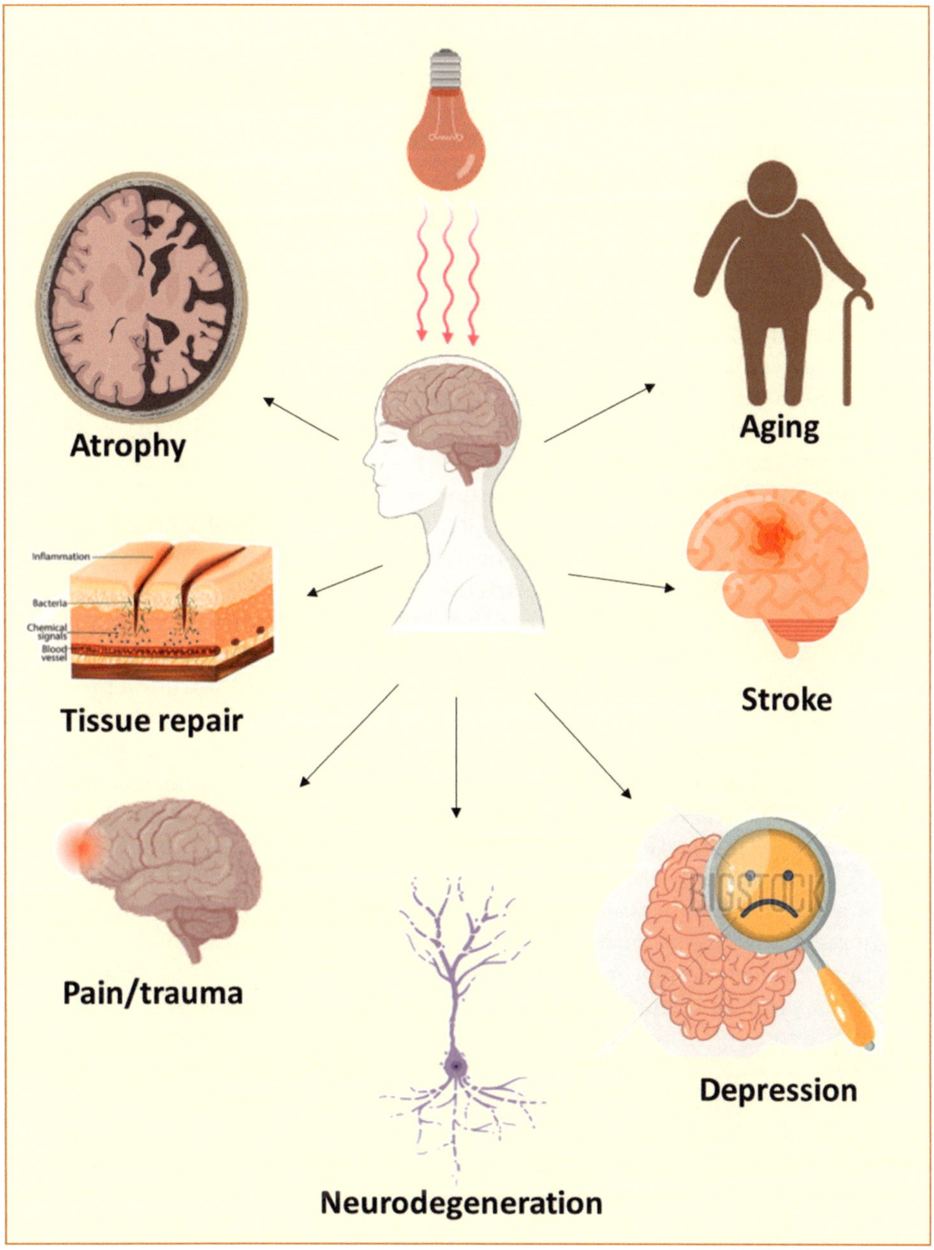

Figure 1.
Beneficial effects of red and near-infrared light on the brain. Diverse reports have shown that irradiation of the brain with red and near-infrared light improves different conditions, including cerebral aging and age-related memory loss, stroke, depression, neurodegeneration in several neurodegenerative diseases, pain and trauma, tissue repair and cicatrization, and atrophy among others.

2. Red light transcranial LED therapy (RL-TCLT): types, devices, uses, and effects

The use of transcranial photobiomodulation is promising in therapeutic and medical benefits for health, with increasing application and projection also in aging and neurodegenerative diseases [9, 24, 25]. The light presents different characteristics that could be used advantageously in the field of health, principally by recent lighting technologies based on an extensive range of diverse light sources, that have been used for photobiomodulation [8]. Light for therapeutic purposes corresponds to a small fraction of the spectrum of luminous radiation, generally in the visible spectrum [8, 25], where it has a biological effect based on the premise of mammalian cellular metabolism from photoreceptors and chromophores molecules [25].

Diverse devices have been used, including Light Amplification by Stimulated Emission of Radiation (Laser) devices, Low-level light laser therapy, and light-emitting diodes (LED) devices [26]. LED devices are semiconductors that present a high efficiency of electrical energy conversion into optical energy, dissipating little thermal energy [26]. Furthermore, these devices can have widely fluctuating power levels depending on the size, number, and power of the individual diodes [16, 27, 28]. LED devices have been compared with lasers; however, devices irradiating LED are bandwidth (approx 40 nm), beam divergence, incoherent radiation emission, and high optical output power; favoring the absorption of energy by different molecular structures [5, 8]. In addition, LED devices have been considered as a safety by the US Food and Drug Administration (FDA) [29].

Red-Light Transcranial Led Therapy (RL-TLTC) involves power-efficient, low heat-producing light sources that have the potential to deliver high-intensity RL of 600-690 wavelengths, that can be pulsed or continuous [30]. In this therapy, the light goes through the layers of the skin and skull, to stimulate the brain and specific cerebral regions, causing biological responses that result in benefits for the individual [7, 8, 31]. In particular, RL-LED mediated a vibrational absorption process, which produces a photochemical effect that leads to the absorption of photons by specific molecules in the cell [5]. In addition, the wavelength (nm), energy density (J/cm^2), and power density (mW/cm^2) are parameters that determine the effectiveness of RL-TLTC. The wavelength of light used is critical since not all ranges of light used have a similar effect, some ranges present reduced effects such as wavelength in the 700–750 nm range. In contrast, RL-LED at 600–690 nm or 760–900 nm has more impact on the biological tissues [5, 8]. Considering that these parameters of light radiation interact with biological tissue, they cause optical phenomena of reflection, transmission, propagation, and absorption. These characteristics also can present variations depending on tissue irradiated, for example by different concentrations of photoreceptor and chromophores molecules that contain biological tissues, like water, cytochromes, and organic molecules as flavins, hemoglobin, and melanin, among others [5]. When light is absorbed, the photon energy reaches the target molecules producing vibrational, rotational, or electronic processes, which generate diverse effects including photochemical, photo-thermal, photo-mechanical, or photo-electrical stimulation [5].

Interestingly, in the use of RL-TLCT, no standard protocol has been established in the literature; moreover, a few reports have shown studies using diverse parameters as varied wavelength ranges, time (sec/min), irradiance, or power density, and energy density with similar results, and important benefits in the brain health [5]. For example, studies applying Transcranial LED therapy bilaterally with wavelengths of 633 and 870 nm, have shown significant progress in both animals models with acute traumatic brain injury, and patients with acute stroke. In both cases also have been observed an improvement in the cognitive capacity post-treatment with this therapy [20]. Other studies using Low-level Laser Therapy (LLLT), with parameters of energy of 3 J/cm^2, a wavelength of 810 nm, and power density of 20 mW/ cm^2, in primary cultured cortical neurons exposed to oxidative stress reveal that LLLT increased the mitochondrial membrane potential and reduced high ROS levels, reducing neuronal death [32]. Similarly, other studies showed that LLLT has a positive impact on neuronal function in both *in vitro* and *in vivo,* enhancing the metabolic capacity of neurons and cognitive functions including memory [14].

Thus, while the transcranial research using RL LED or laser remains in the initial stages, growing evidence showed that although the RL and NIR-light therapy presents a wide range of characters, can modulate cell activity, including energy metabolism and cell function [25]. This is relevant since these therapies can lead to the improvement of pathological conditions and be in future significant clinical contribution, for example performing clinical treatments that allow helping older persons to prevent or mitigate the age-related cognitive impairment (**Figure 1**).

3. Effects of RL-TCLT on hippocampal structure and function

The effects of RL-TCLT on the nervous system, neuronal repair, and improving cognition are growing, and have been well-documented in cellular, animal models, and human studies [8]. Since the past decade, the use of RL-TCLT as an advanced and non-invasive therapeutic method in several brain-related conditions has attracted interest from researchers in biomedical science, including those conditions or pathologies that manifest memory loss. Nevertheless, the underlying neural mechanisms are not well understood. The hippocampus is a brain structure of special importance in studying aging and cognitive decline, its main function is learning and memory [33]. It is a dorsoventrally elongated area, composed of the dentate gyrus (DG), the cornu ammonis (CA) fields CA1, CA2, and CA3, and the subiculum cortex [34]. The trisynaptic circuit is the main excitatory hippocampal synaptic pathway, formed by 3 neuronal groups: granule cells in the DG, and pyramidal neurons of the CA1 and CA3 [35]. This circuit receives inputs from the superficial layers of the entorhinal cortex via the perforant path to the DG. The DG projects to the CA3, which in turn projects to the CA1. Thus, CA1 projects to the deep layers of the entorhinal cortex, closing the circuit [33, 35, 36]. The hippocampus mediates recognition and spatial memory, by a highly regulated circuit with a high-energy demand [37]. Besides, is important to highlight that the hippocampus is highly susceptible to factors such as mitochondrial dysfunction, stress, inflammation, or physiological process such as aging, accumulating damage that gradually lead to a loss of hippocampal function [33].

Spatial memory gradually decreases with the age, since the hippocampus is critical for this type of memory, and the impairment of hippocampal neurons unequivocally results in spatial memory diminishing [37]. Studies from our and other groups have shown the reduced capacity of aged mice to learn and remember spatial tasks [4, 38, 39]. This is indicated by increased time to find a hidden platform in the Morris Water Maze (MWM) or a hidden chamber in the Barnes Maze (BM), two classic probes to evaluate hippocampus-dependent spatial memory [4, 22]. A report using the senescence-accelerated prone 8 (SAMP8) mice, a mouse model widely used to study oxidative impairment, and age-related brain damage, showed that RL-TCLT at 630 nm for two consecutive months prevents spatial memory loss in 5 month-old (mo) SAMP8 mice, and more importantly rescued the cognitive deficits in SAMP8 mice of 7 mo [40]. This last was accompanied by reduced ROS levels in the brain and increased activity of antioxidant enzymes such as catalase and formaldehyde dehydrogenase [40]. Similarly, TCLT with NIR laser at 810 nm applied in mice exposed to acute sleep deprivation showed reduced hippocampal oxidative damage, increasing the activity of antioxidant enzymes, including superoxide dismutase (SOD) and glutathione peroxidase (GPx) [16]. Additionally, several studies with cells, animals and in clinical trial conclude that RL-TCLT may have a potential effect on the brain since that has been observed that RL-TCTL protect nerve cells from a future impairment, reducing permanent neuronal damage and increasing their survival. For example, the treatment of K369I tau (K3) mice, a transgenic mouse model of tauopathies and Alzheimer's Disease, with NIR (600–1000 nm) 20 times four weeks reveal a reduction in the size and number of amyloid-β plaques in the neocortex and hippocampus [41].

Besides, RL-TCLT may have a potential effect promoting both synaptogenesis and neurogenesis [42]. Both processes are essential to facilitating connectivity, neural regeneration, and generate structural changes that help to maintain existing neurons, and to encourage the growth of new neurons and synapses process [43, 44]. In this context, IR-light at 808 nm (350 mW/cm2 and 294 J) was applied in the scalp of a photothrombotic model of ischemic stroke in rats for seven days during 2-minute daily. The authors observed that IR-light therapy significantly attenuated behavioral deficits and infarct volume in cortical regions induced by photothrombotic stroke. This improvement was accompanied by neurogenesis and synaptogenesis, as is indicated by increased immunoreactivity of the proliferative and differentiation markers BrdU, Ki67, DCX, MAP2, spinophilin, and the synaptic marker synaptophysin [45]. Also, other clinical studies reveal positive effects of transcranial LED therapy on cerebral blood fluid (CBF) in patients in a vegetative state or with major depression and anxiety. LED treatment by 20 or 30 min per session, thrice per week over 6 weeks, or two times daily for over seventy days, with different wavelengths of 610, 627, and 810 nm increase CBF, improving cerebral vascular perfusion and reducing brain disorders [9, 46, 47]. Similarly, the application of TC-LLL therapy at 810-nm in mice model of cortical impact and traumatic brain injury reveal increased proliferating neural cells around the lesion, possibly activating regenerative mechanisms such as inducing neurogenesis in the dentate gyrus of the hippocampus [48]. Besides, they observed that the mice treated improved learning and memory reducing cognitive impairment [48, 49].

Thus, while its positive effects have been demonstrated countless times in animal models, they have yet to be proven in broad-scope clinical testing. However, the research that does exist is very promising, strongly indicating that RL-TCLT could be a viable treatment for a broad range of neurological diseases including stroke, traumatic brain injury, Parkinson's disease, Alzheimer's disease, and depression, in addition to providing cognitive enhancement for healthy subjects of advanced age that manifest cognitive impairment.

4. Effects of RL-TCLT on synaptic neurotransmission and synaptic plasticity

Considering that irradiation with RL between 600 and 1200 nm produces changes at molecular, cellular, and tissue levels [50, 51] improving cognitive capacities [52], this enhancement in brain function will result in synaptic neurotransmission and synaptic plasticity potentiation after light treatment [53, 54]. Neurotransmission and synaptic plasticity represent the capacity of synaptic connections to adapt structurally and functionally in a stimulus-dependent manner [43]. Both synaptic neurotransmission and synaptic plasticity can be affected by different factors, such as mitochondrial dysfunction and increased oxidative stress, as well as physiological events including aging, stroke, brain injuries, or neurodegenerative disease, among others [22, 55, 56]. Therefore, treatments focused in maintain or promote neurotransmission and synaptic plasticity are attracting increasing attention. In this context, despite the beneficial effects showed for RL-TCLT on cognition, practically not exist electrophysiological studies using this therapy. As an approximation, we will discuss the studies using transcranial low-level laser light (TC-LLL).

Studies both *in vitro* and *in vivo* have shown that TC-LLL therapy supports neural function, this has been observed principally in reports using transgenic mouse models of Alzheimer's disease (AD) [57]. Meng et al. observed that TC-LLL therapy at 632.8 nm in primary hippocampal neurons treated with full-length $A\beta_{1-42}$ peptide reduced Aβ-induced neurotoxicity. In addition, TC-LLL therapy shows neuroprotective effects decreasing Aβ-induced dendrite atrophy [57]. Also, TC-LLL treatment increased the expression of brain-derived neurotrophic factor (BDNF) in cell line and cultured neurons derived from APP/PS1 transgenic mice, suggesting that this neurotrophin will be modulating dendritic structure, promoting the survival of neurons and dendrite growth, and potentiating synaptic transmission in the CNS [57, 58]. All these results can be explained by the activation of the ERK/CREB/BDNF pathway mediated by TC-LLL therapy [45] because this pathway is involved in the dendritic development of neurons [45, 57, 59]. Therefore, the TC-LLL therapy can induce activation of signaling transduction pathways, and gene transcription, which increases protein expression of different synaptic effectors and modulators, effects that also are potential therapeutic in treating neurodegenerative disease.

Interestingly, the NIR-LED light treatment at 670 nm in the Tg2576 mice model of AD, which progressively accumulated Aβ in their brain [60], indicate that NIR-LED therapy decreased the levels of $A\beta_{1-42}$ at the synapses and Aβ oligomer-induced reduction in long-term potentiation (LTP), relevant processes of neuroplasticity that correlates with memory formation [61]. Therefore, NIR-LED light therapy recovered crucial processes related to synaptic function, necessary to the preservation of cognition abilities [62]. Additionally, studies with photobiomodulation transcranial therapy with wavelengths of 635 nm in a mouse model of depression showed that this treatment reduces glutamate levels and neurotoxicity, improving the depressant behavior. These beneficial effects can be explained by the activation of the PKA pathway and the increased levels of the α-amino-3-hydroxy-5-methyl-4-isoxazolepropionic acid (AMPA) receptors. In addition, reduce the expression of GluA1, decreasing the glutamatergic neurotransmission. Thus this therapy could rescue excitatory synaptic transmission, improve synaptic plasticity, and have also a potential anti-depressive effect [63].

Thus, several pieces of evidence suggest that the application of transcranial therapy light could be used to improve cellular components associated with the synaptic function [42], which is essential in the maintenance and preservation of cognition, including learning and memory [43]. In addition to the therapeutic effects at the molecular level, it is proposed the generation of changes at the behavioral level, such as cognitive improvement, antidepressant effects, and sleep improvement [42]. Furthermore, this therapy can stimulate neuronal organization or reorganization, therefore it could be extremely promising as a method of stabilization and/or improvement of various brain disorders or nervous system, and neurodegenerative diseases [30, 42]. However, more extensive studies are necessary to evidence all the cellular and molecular mechanisms involved in these encouraging results. This last especially considering that the evidence summarized here consider severe differences in the device and light type used, the protocol of administration, and the study model. Is imperative to advance understanding the multiple targets of red light in the synaptic structure and function.

5. RL-TCLT mechanisms: improving mitochondrial function

Considering the multiple reports revised previously, now known that Red Light therapy, including Red and InfraRed LED light and Red Laser light, have favorable effects on brain structure and function, and especially in the hippocampus improving cognitive functions [9, 25, 42]. This enhances in cognitive capacity could be explained by the activation of neurogenesis and synaptogenesis [9, 42], as well as by the stimulation of processes related to synaptic plasticity such as LTP [62]. However, any of these events reveal a potential mechanism by which Red light treatment results therapeutic to different affections such as aging, neurodegenerative disease, stroke, and depression among others [42].

Interestingly, all pathological conditions mentioned previously involve, almost in part, dysfunction of hippocampal neurons attributable to mitochondrial defects [64, 65]. For example, mitochondrial dysfunction is considered a hallmark of aging and could be considered one of the factors leading to neurodegeneration [4, 22, 66]. Studies in humans and animal models showed that decreased memory correlates with reduced cerebral energetics metabolism and more specifically to mitochondrial bioenergetics deficits [4, 22, 66]. Therefore, the mitochondrial focus of aging and neurodegenerative diseases is of great interest for the development of a potent and ideally non-invasive anti-aging intervention to improve or attenuate cognitive impairment in the elderly.

Notably, enhanced metabolic functioning is one of the most identifiable properties of irradiate neuronal cells with RL or NIR light, resulting in increased intracellular ATP production [8]. Thus, mitochondrial ATP production is one of the most strongly suggested mechanisms of action of RL therapy [5, 8]; for example, studies using RL-TCLT at 660 nm for 15 sec daily for 2 weeks in aged 18 mo mice improved ATP concentration [16]. More specifically, studies in vitro with RL and NRL LED radiations with a wavelength between 600 and 850 nm have shown that the effects of this treatment are principally attributed to photon absorption by complex IV of the mitochondrial respiratory chain [5]. This mitochondrial complex corresponds to the cytochrome c oxidase (COX) enzyme [22] and it seems that RL increases the activity of this enzymatic complex, leading to enhancement of oxygen consumption and ultimately to mitochondrial respiration [5]. COX is a photo acceptor of RL and NIR light, which generates a redox change in the enzyme [5, 8]. In turn, this causes a transient change in mitochondrial membrane potential (mψ) and increases ATP production [5, 16]. Thus, wavelengths corresponding or near to red will be improving the mitochondrial production of ATP, potentiating the synaptic and cognitive function [5, 42]. Nevertheless, is important to highlight that other works report that RL could inhibit the COX enzyme. In particular, NIR wavelengths of 750 nm and 950 nm reduced the activity of the COX complex. This results in decreased mitochondrial respiration and a loss of mitochondrial membrane potential ($\Delta\Psi$m) [67]. Is surprising to note that the attenuation of mitochondrial function and the concomitant production of superoxide radical reduce neuronal death exposed to oxygen–glucose deprivation and in a mice model of ischemia, an effect that is not observed after other NIR wavelengths that activate COX [67]. Altogether, these contradictory results question the real effect of RL on COX responsible for the beneficial effects of this therapy (**Figure 2**).

On the other hand, several reports showed that RL-LED modulates the levels of reactive oxygen species (ROS) [30, 40]. Studies using RL-LED illumination at 630 nm reduces brain H_2O_2 levels in cultured cells and the brain of SAMP8 mice [40]. This could be explained by an increment in the activity of antioxidant enzymes such as catalase or also could be a consequence of increased mitochondrial function with reduced electron leak [5, 16]; more studies are necessary to evaluate these possibilities. Besides, in this study, the authors showed that RL-LED absorption activates transcription factors that regulate long-lasting effects on gene expression [16], therefore this suggests that Red Light therapy could be a more complex mechanism, at a long time, and not only a transient activation of several enzymes. Other results also showed that Red 635 nm irradiation inhibits the expression of COX

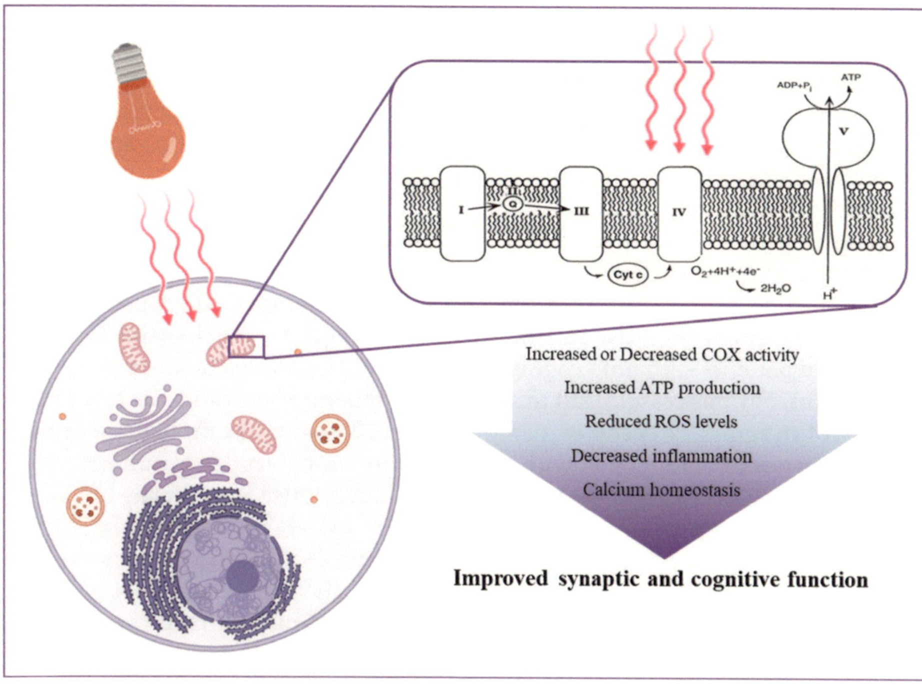

Figure 2.
Mechanism of action proposed to red and near-infrared light and its cellular effects. Transcranial therapy using red and near-infrared light has been proposed to photoactivate the cytochrome c oxidase (COX) enzyme, the complex IV of the electron transporter chain of the mitochondria. However other reports propose that several wavelengths inhibit COX enzyme; modulating ROS and ATP production, calcium homeostasis, and inflammatory processes.

enzyme, reducing ROS levels and mRNA of cytosolic phospholipase A2 (cPLA2) and secretary phospholipase A2 (sPLA2) [68]. This also consequently inhibits the release of PGE2, suggesting an additional anti-inflammatory effect (**Figure 2**).

Additional mechanisms that will be involved in the positive effects of RL-LED implicate Ca^{2+} ions modulation [8, 40]. RL and NIR LED are recognized by water groups formed in the heat/light-gated Ca^{+2} channel. This induces vibrational water energy, which in turn disorganizes the protein structure of the Ca^{+2} channel. This conformational change finally leads to channel opening; modulating intracellular Ca^{+2} levels [5]. This possible mechanism is relevant in neurons, considering that intracellular Ca^{+2} levels are critical to trigger survival or death pathways related to synaptic activity [69].

In summary, despite various mechanisms that could be mentioned such as the potential molecular target of RL and NRL, still is necessary additional research in the field to understand the events that result in synaptic and cognitive function. Possibly these improvements are the result of diverse events occurring simultaneously.

6. Future perspectives of transcranial Red630-light-transcranial LED therapy preventing age-related memory loss: Our advances

Despite diverse studies shown possible molecular targets of RL-LED therapy [5, 8], the precise mechanism underlying the neuroprotective actions of RL-TCLT is not completely understood. Therefore, more studies are required to determine the biological events that lead to neuroprotection or neuronal repair in both aging and neurodegenerative diseases. Possibly, the main problem related to the incapacity of determining a detailed mechanism is based on the variability of wavelengths, times of treatment, and models used [15, 16, 40, 63, 70]. Therefore, this highlights the need for complete studies using the same mice model, LED dispositive, and therapy protocol, to understand and describe the mechanism(s) underlying the benefits of RL-TCLT.

Interestingly, RL-TCLT at 630 nm in patients with traumatic brain injuries, using a helmet that emits radiation for 30 min, three times per week, for six weeks showed a great reduction in post-traumatic stress symptoms, insomnia, and depression, suggesting improved cognitive function [71]. More importantly, the same RL-TCLT used in aged patients with mild cognitive impairment improves memory in these aged humans. For this reason, and to study the complete effects and mechanisms of RL-TCLT in aging, we designed a unique RL-TCLT device to emit homogeneous light at a wavelength of 630 nm, with 100 J of energy, a power density of 0,35 w/cm^2, and an energy density of 43.5 J/cm^2 in the brain of mice, specifically in the hippocampus.

We applied RL-TCLT to the hippocampus of 7.5mo SAMP8 mice, a mice model of accelerated aging, with an irradiation time of 125 s daily (excluding weekends) for 5 weeks. This protocol is equivalent to the applied to patients with mild cognitive impairment described previously, and the mouse lifespan. We started the RL-TCLT in SAMP8 at 7.5mo because we and other authors showed that the non-transgenic SAMP8 mice present age-related hippocampal memory loss since 6mo and is more evident from 7mo onwards [72]. Interestingly, our results reveal that 7.5mo SAMP8 mice treated with RL-TCLT at the hippocampus improves spatial learning and memory of aged SAMP8 mice. This cognitive improvement will be due to a possible remodeling of the synaptic structure toward more active synapses reducing the risk of excitotoxic events. This is suggested by i) an increase in presynaptic proteins such as synaptophysin (SYP) and Synapsin (SYN) that increase the neurotransmitter release [73], ii) a decrease in the NMDAR subunit NR2B, whose protein levels are related to excitotoxicity [74] and iii) higher Arc protein levels, a marker of synaptic plasticity [75] (Jara et al., manuscript in preparation).

Considering that both memory formation and synaptic activity are highly dependent on energy [76], that mitochondria are the main ATP producer of the cell [22], and that the suggested mechanisms by RL-TCLT target the mitochondria [5], we evaluated different mitochondrial functions in the hippocampus of treated SAMP8 mice with RL-TCLT. Relevantly, we observed increased ATP production, higher activity of the OXPHOS complex II-III, and IV (COX enzyme); suggesting that RL-TCLT directly stimulates mitochondrial bioenergetics function enhancing the activity of other OXPHOS complexes in addition to COX (Jara et al., manuscript in preparation). Similarly, we observed decreased levels of the mitochondrial calcium uniporter (MCU), suggesting that it will result in reduced mitochondrial Ca^{+2} overload and swelling, enhancing mitochondrial Ca^{+2} buffering. In fact, this last was validated in Ca^{+2} overload assays in hippocampal mitochondria from RL-TCLT SAMP8 mice (Jara et al., manuscript in preparation), indicating that RL-TCLT also improves the calcium buffering capacity of the aged hippocampal mitochondria. Whether bioenergetics and calcium buffering enhancing are directly related or are independent mechanisms requires future analysis.

Thus, our results indicated RL-LED-mediated mitochondrial stimulation, which could be transient or permanent. But it is difficult to think that only transient activation of mitochondrial function could explain the improved cognitive effects produced by RL-TCLT treatment. Although mitochondrial ATP production is vital for synaptic communication, it is probably not solely sufficient to result in improved hippocampal memory. Therefore, is highly probable that other mechanisms are involved in the beneficial effects of RL-TCLT, which result in gene transcription and the consequent cellular remodeling. In concordance with the anterior, we also observed higher PGC-1α protein levels, a transcriptional coactivator considered the main inducer of mitochondrial biogenesis that also regulates mitochondrial function [77]. This suggests that RL-TCLT will stimulate the generation of new mitochondria or the activation of gene-dependent mitochondrial reparation pathways that result in increased mitochondrial function. However, this requires a robust study.

7. Conclusions

In conclusion, despite the majority of treatments using RL-LED therapy are focused on cosmetic applications, RL-TCLT generates cellular effects that will be used to treat different affections including aging and neurodegenerative disease. For this, is necessary to be prudent to decide the more adequate wavelength, light intensity, and duration of therapy, because according to these parameters will found positive or negative effects. In addition, RL-TCLT seems to improve antioxidant defenses and mitochondrial function by enhancing COX IV activity. In concordance with these findings, we also propose that RL-TCLT stimulates mitochondrial function, both enhancing OXPHOS-mediated mitochondrial bioenergetics and calcium buffering capacity. Finally, we suggest that prolonged exposition to RL-TCLT result in a permanent remodeling of the cell, by a mechanism that involves gene transcription, which results in higher synaptic and cognitive function. Future studies are extremely necessary to solve all the questions regarding the benefits of RL-TCLT.

Acknowledgements

This work was supported by FONDECYT N° 3210591 to CJ, FONDECYT N°11170546, and CONICYT PAI N°77170091 to CTR.

Conflict of interest

The authors declare no conflict of interest.

Author details

Claudia Jara[1†], Débora Buendía[2,3,4†], Alvaro Ardiles[4], Pablo Muñoz[4] and Cheril Tapia-Rojas[1*]

1 Laboratory of Neurobiology of Aging, Centro de Biología Celular y Biomedicina (CEBICEM), Facultad de Medicina y Ciencia, Universidad San Sebastián, Sede Los Leones, Santiago, Chile

2 Escuela de Ingeniería Civil Biomédica de la Universidad de Valparaíso, Valparaíso, Chile

3 Universidade Anhembi Morumbi. Sao Paulo, Brasil

4 Translational Neurology Center, Faculty of Medicine, Biomedical Research Center, Universidad de Valparaíso, Valparaíso, Chile

*Address all correspondence to: cheril.tapia@uss.cl

† Both authors contributed equally.

References

[1] Peters, R., *Ageing and the brain.* Postgrad Med J, 2006. **82**(964): p. 84-88.

[2] Jackow-Nowicka, J., et al., *The Impact of Common Epidemiological Factors on Gray and White Matter Volumes in Magnetic Resonance Imaging-Is Prevention of Brain Degeneration Possible?* Front Neurol, 2021. **12**: p. 633619.

[3] Preston, A.R. and H. Eichenbaum, *Interplay of hippocampus and prefrontal cortex in memory.* Curr Biol, 2013. **23**(17): p. R764-R773.

[4] Olesen, M.A., et al., *Premature synaptic mitochondrial dysfunction in the hippocampus during aging contributes to memory loss.* Redox Biol, 2020. **34**: p. 101558.

[5] Hamblin, M.R., *Mechanisms and Mitochondrial Redox Signaling in Photobiomodulation.* Photochem Photobiol, 2018. **94**(2): p. 199-212.

[6] Khan, I. and P.R. Arany, *Photobiomodulation Therapy Promotes Expansion of Epithelial Colony Forming Units.* Photomed Laser Surg, 2016. **34**(11): p. 550-555.

[7] Avci, P., et al., *Low-level laser (light) therapy (LLLT) in skin: stimulating, healing, restoring.* Semin Cutan Med Surg, 2013. **32**(1): p. 41-52.

[8] de Freitas, L.F. and M.R. Hamblin, *Proposed Mechanisms of Photobiomodulation or Low-Level Light Therapy.* IEEE J Sel Top Quantum Electron, 2016. **22**(3).

[9] Salehpour, F., et al., *Brain Photobiomodulation Therapy: a Narrative Review.* Mol Neurobiol, 2018. **55**(8): p. 6601-6636.

[10] Pitzschke, A., et al., *Red and NIR light dosimetry in the human deep brain.* Phys Med Biol, 2015. **60**(7): p. 2921-2937.

[11] Haeussinger, F.B., et al., *Simulation of near-infrared light absorption considering individual head and prefrontal cortex anatomy: implications for optical neuroimaging.* PLoS One, 2011. **6**(10): p. e26377.

[12] Naeser, M.A., et al., *Transcranial, Red/Near-Infrared Light-Emitting Diode Therapy to Improve Cognition in Chronic Traumatic Brain Injury.* Photomed Laser Surg, 2016. **34**(12): p. 610-626.

[13] Dungel, P., et al., *Low level light therapy by LED of different wavelength induces angiogenesis and improves ischemic wound healing.* Lasers Surg Med, 2014. **46**(10): p. 773-780.

[14] Rojas, J.C. and F. Gonzalez-Lima, *Neurological and psychological applications of transcranial lasers and LEDs.* Biochem Pharmacol, 2013. **86**(4): p. 447-457.

[15] Naderi, M.S., et al., *A Comparative Study of 660 nm Low-Level Laser and Light Emitted Diode in Proliferative Effects of Fibroblast Cells.* J Lasers Med Sci, 2017. **8**(Suppl 1): p. S46-S50.

[16] Salehpour, F., et al., *A Protocol for Transcranial Photobiomodulation Therapy in Mice.* J Vis Exp, 2018(141).

[17] Sharma, S.K., et al., *Dose response effects of 810 nm laser light on mouse primary cortical neurons.* Lasers Surg Med, 2011. **43**(8): p. 851-859.

[18] Salehpour, F., et al., *Transcranial low-level laser therapy improves brain mitochondrial function and cognitive impairment in D-galactose-induced aging mice.* Neurobiol Aging, 2017. **58**: p. 140-150.

[19] Yue, X., et al., *New insight into Alzheimer's disease: Light reverses*

Abeta-obstructed interstitial fluid flow and ameliorates memory decline in APP/PS1 mice. Alzheimers Dement (N Y), 2019. **5**: p. 671-684.

[20] Naeser, M.A., et al., *Improved cognitive function after transcranial, light-emitting diode treatments in chronic, traumatic brain injury: two case reports.* Photomed Laser Surg, 2011. **29**(5): p. 351-358.

[21] Naeser, M.A. and M.R. Hamblin, *Potential for transcranial laser or LED therapy to treat stroke, traumatic brain injury, and neurodegenerative disease.* Photomed Laser Surg, 2011. **29**(7): p. 443-446.

[22] Jara C. , T.K.A., Olesen A. M and Cheril Tapia-Rojas, *Mitochondrial Dysfunction as a Key Event during Aging: From Synaptic Failure to Memory Loss*, in *Mitochondria and Brain Disorders*, S. Baloyannis, Editor. 2020, IntechOpen: London. p. 387-411.

[23] Wong-Riley, M.T., et al., *Photobiomodulation directly benefits primary neurons functionally inactivated by toxins: role of cytochrome c oxidase.* J Biol Chem, 2005. **280**(6): p. 4761-4771.

[24] Salehpour, F. and S.H. Rasta, *The potential of transcranial photobiomodulation therapy for treatment of major depressive disorder.* Rev Neurosci, 2017. **28**(4): p. 441-453.

[25] Hamblin, M.R., *Photobiomodulation for Alzheimer's Disease: Has the Light Dawned?* Photonics, 2019. **6**(3).

[26] Heiskanen, V. and M.R. Hamblin, *Photobiomodulation: lasers vs. light emitting diodes?* Photochem Photobiol Sci, 2018. **17**(8): p. 1003-1017.

[27] Gutierrez-Menendez, A., et al., *Photobiomodulation as a promising new tool in the management of psychological disorders: A systematic review.* Neurosci Biobehav Rev, 2020. **119**: p. 242-254.

[28] Santos, J., et al., *Effects of transcranial LED therapy on the cognitive rehabilitation for diffuse axonal injury due to severe acute traumatic brain injury: study protocol for a randomized controlled trial.* Trials, 2018. **19**(1): p. 249.

[29] Dodd, E.M., et al., *Photobiomodulation therapy for androgenetic alopecia: A clinician's guide to home-use devices cleared by the Federal Drug Administration.* J Cosmet Laser Ther, 2018. **20**(3): p. 159-167.

[30] Hamblin, M.R., *Shining light on the head: Photobiomodulation for brain disorders.* BBA Clin, 2016. **6**: p. 113-124.

[31] Huang, Y.Y., et al., *Transcranial low level laser (light) therapy for traumatic brain injury.* J Biophotonics, 2012. **5**(11-12): p. 827-837.

[32] Huang, Y.Y., et al., *Low-level laser therapy (LLLT) reduces oxidative stress in primary cortical neurons in vitro.* J Biophotonics, 2013. **6**(10): p. 829-838.

[33] Lazarov, O. and C. Hollands, *Hippocampal neurogenesis: Learning to remember.* Prog Neurobiol, 2016. **138-140**: p. 1-18.

[34] Stepan, J., J. Dine, and M. Eder, *Functional optical probing of the hippocampal trisynaptic circuit in vitro: network dynamics, filter properties, and polysynaptic induction of CA1 LTP.* Front Neurosci, 2015. **9**: p. 160.

[35] Witter, M.P., et al., *Architecture of spatial circuits in the hippocampal region.* Philos Trans R Soc Lond B Biol Sci, 2014. **369**(1635): p. 20120515.

[36] Hartley, T., et al., *Space in the brain: how the hippocampal formation supports spatial cognition.* Philos Trans R Soc Lond B Biol Sci, 2014. **369**(1635): p. 20120510.

[37] Anand, K.S. and V. Dhikav, *Hippocampus in health and disease: An*

overview. Ann Indian Acad Neurol, 2012. **15**(4): p. 239-246.

[38] Leal, S.L. and M.A. Yassa, *Neurocognitive Aging and the Hippocampus across Species.* Trends Neurosci, 2015. **38**(12): p. 800-812.

[39] Jara, C., et al., *Tau Deletion Prevents Cognitive Impairment and Mitochondrial Dysfunction Age Associated by a Mechanism Dependent on Cyclophilin-D.* Front Neurosci, 2020. **14**: p. 586710.

[40] Zhang, J., et al., *Illumination with 630 nm Red Light Reduces Oxidative Stress and Restores Memory by Photo-Activating Catalase and Formaldehyde Dehydrogenase in SAMP8 Mice.* Antioxid Redox Signal, 2019. **30**(11): p. 1432-1449.

[41] Berman, M.H., et al., *Photobiomodulation with Near Infrared Light Helmet in a Pilot, Placebo Controlled Clinical Trial in Dementia Patients Testing Memory and Cognition.* J Neurol Neurosci, 2017. **8**(1).

[42] Hennessy, M. and M.R. Hamblin, *Photobiomodulation and the brain: a new paradigm.* J Opt, 2017. **19**(1): p. 013003.

[43] Mateos-Aparicio, P. and A. Rodriguez-Moreno, *The Impact of Studying Brain Plasticity.* Front Cell Neurosci, 2019. **13**: p. 66.

[44] Abraham, W.C., O.D. Jones, and D.L. Glanzman, *Is plasticity of synapses the mechanism of long-term memory storage?* NPJ Sci Learn, 2019. **4**: p. 9.

[45] Yan, X., et al., *Low-level laser irradiation modulates brain-derived neurotrophic factor mRNA transcription through calcium-dependent activation of the ERK/CREB pathway.* Lasers Med Sci, 2017. **32**(1): p. 169-180.

[46] Nawashiro, H., et al., *Focal increase in cerebral blood flow after treatment with near-infrared light to the forehead in a patient in a persistent vegetative state.* Photomed Laser Surg, 2012. **30**(4): p. 231-233.

[47] Hipskind, S.G., et al., *Pulsed Transcranial Red/Near-Infrared Light Therapy Using Light-Emitting Diodes Improves Cerebral Blood Flow and Cognitive Function in Veterans with Chronic Traumatic Brain Injury: A Case Series.* Photobiomodul Photomed Laser Surg, 2019. **37**(2): p. 77-84.

[48] Xuan, W., et al., *Transcranial low-level laser therapy enhances learning, memory, and neuroprogenitor cells after traumatic brain injury in mice.* J Biomed Opt, 2014. **19**(10): p. 108003.

[49] Xuan, W., et al., *Transcranial low-level laser therapy improves neurological performance in traumatic brain injury in mice: effect of treatment repetition regimen.* PLoS One, 2013. **8**(1): p. e53454.

[50] Wang, Y., et al., *Photobiomodulation of human adipose-derived stem cells using 810nm and 980nm lasers operates via different mechanisms of action.* Biochim Biophys Acta Gen Subj, 2017. **1861**(2): p. 441-449.

[51] Chung, H., et al., *The nuts and bolts of low-level laser (light) therapy.* Ann Biomed Eng, 2012. **40**(2): p. 516-533.

[52] Barrett, D.W. and F. Gonzalez-Lima, *Transcranial infrared laser stimulation produces beneficial cognitive and emotional effects in humans.* Neuroscience, 2013. **230**: p. 13-23.

[53] Li, Z., et al., *The importance of dendritic mitochondria in the morphogenesis and plasticity of spines and synapses.* Cell, 2004. **119**(6): p. 873-887.

[54] Massaad, C.A. and E. Klann, *Reactive oxygen species in the regulation of synaptic plasticity and memory.* Antioxid Redox Signal, 2011. **14**(10): p. 2013-2054.

[55] Jha, S.K., et al., *Stress-Induced Synaptic Dysfunction and Neurotransmitter Release in Alzheimer's Disease: Can Neurotransmitters and Neuromodulators be Potential Therapeutic Targets?* J Alzheimers Dis, 2017. **57**(4): p. 1017-1039.

[56] Li, N., et al., *Evidence for impaired plasticity after traumatic brain injury in the developing brain.* J Neurotrauma, 2014. **31**(4): p. 395-403.

[57] Meng, C., Z. He, and D. Xing, *Low-level laser therapy rescues dendrite atrophy via upregulating BDNF expression: implications for Alzheimer's disease.* J Neurosci, 2013. **33**(33): p. 13505-13517.

[58] Murer, M.G., Q. Yan, and R. Raisman-Vozari, *Brain-derived neurotrophic factor in the control human brain, and in Alzheimer's disease and Parkinson's disease.* Prog Neurobiol, 2001. **63**(1): p. 71-124.

[59] Kwon, M., et al., *BDNF-promoted increases in proximal dendrites occur via CREB-dependent transcriptional regulation of cypin.* J Neurosci, 2011. **31**(26): p. 9735-9745.

[60] Westerman, M.A., et al., *The relationship between Abeta and memory in the Tg2576 mouse model of Alzheimer's disease.* J Neurosci, 2002. **22**(5): p. 1858-1867.

[61] Ahnaou, A., et al., *Aging Alters Olfactory Bulb Network Oscillations and Connectivity: Relevance for Aging-Related Neurodegeneration Studies.* Neural Plast, 2020. **2020**: p. 1703969.

[62] Comerota, M.M., B. Krishnan, and G. Taglialatela, *Near infrared light decreases synaptic vulnerability to amyloid beta oligomers.* Sci Rep, 2017. **7**(1): p. 15012.

[63] Zhang, D., et al., *Photobiomodulation Therapy Ameliorates Glutamatergic Dysfunction in Mice with Chronic Unpredictable Mild Stress-Induced Depression.* Oxidative Medicine and Cellular Longevity, 2021. **2021**: p. 6678276.

[64] Hiebert, J.B., et al., *Traumatic brain injury and mitochondrial dysfunction.* Am J Med Sci, 2015. **350**(2): p. 132-138.

[65] Huang, W.J., X. Zhang, and W.W. Chen, *Role of oxidative stress in Alzheimer's disease.* Biomed Rep, 2016. **4**(5): p. 519-522.

[66] Torres, A.K., et al., *Pathologically phosphorylated tau at S396/404 (PHF-1) is accumulated inside of hippocampal synaptic mitochondria of aged Wild-type mice.* Sci Rep, 2021. **11**(1): p. 4448.

[67] Sanderson, T.H., et al., *Inhibitory modulation of cytochrome c oxidase activity with specific near-infrared light wavelengths attenuates brain ischemia/ reperfusion injury.* Sci Rep, 2018. **8**(1): p. 3481.

[68] Lim, W., et al., *The anti-inflammatory mechanism of 635 nm light-emitting-diode irradiation compared with existing COX inhibitors.* Lasers Surg Med, 2007. **39**(7): p. 614-621.

[69] Ureshino, R.P., et al., *The Interplay between Ca(2+) Signaling Pathways and Neurodegeneration.* Int J Mol Sci, 2019. **20**(23).

[70] Xuan, W., et al., *Low-level laser therapy for traumatic brain injury in mice increases brain derived neurotrophic factor (BDNF) and synaptogenesis.* J Biophotonics, 2015. **8**(6): p. 502-511.

[71] Carneiro, A.M.C., et al., *Transcranial Photobiomodulation Therapy in the Cognitive Rehabilitation of Patients with Cranioencephalic Trauma.* Photobiomodul Photomed Laser Surg, 2019. **37**(10): p. 657-666.

[72] Liu, B., J. Liu, and J.S. Shi, *SAMP8 Mice as a Model of Age-Related Cognition*

Decline with Underlying Mechanisms in Alzheimer's Disease. J Alzheimers Dis, 2020. **75**(2): p. 385-395.

[73] Hilfiker, S., et al., *Synapsins as regulators of neurotransmitter release.* Philos Trans R Soc Lond B Biol Sci, 1999. **354**(1381): p. 269-279.

[74] Jakaria, M., et al., *Neurotoxic Agent-Induced Injury in Neurodegenerative Disease Model: Focus on Involvement of Glutamate Receptors.* Front Mol Neurosci, 2018. **11**: p. 307.

[75] Korb, E. and S. Finkbeiner, *Arc in synaptic plasticity: from gene to behavior.* Trends Neurosci, 2011. **34**(11): p. 591-598.

[76] Kann, O. and R. Kovacs, *Mitochondria and neuronal activity.* Am J Physiol Cell Physiol, 2007. **292**(2): p. C641-C657.

[77] Gureev, A.P., E.A. Shaforostova, and V.N. Popov, *Regulation of Mitochondrial Biogenesis as a Way for Active Longevity: Interaction Between the Nrf2 and PGC-1alpha Signaling Pathways.* Front Genet, 2019. **10**: p. 435.

Chapter 9

New Prospects for Stem Cell Therapy in Alzheimer's Disease

Kun Jiang, Yongqi Zhu and Lei Zhang

Abstract

Alzheimer's disease (AD) is a kind of neurodegenerative disease with insidious onset and progressive progression. The etiology of AD may be related to the loss of neurons, astrocytes, and microglial in the nervous system. Exogenous stem cell transplantation has brought hope to the treatment of AD. Stem cell transplantation can reduce amyloid β-protein (Aβ) deposition and Tau phosphorylation, and provide secretory factor support to improve learning and memory deficits. The purpose of this review is to provide an overview of the relationship between different stem cell species and the treatment of AD, and also summarize current experimental stem cell therapy strategies and their potential clinical applications in the future.

Keywords: Stem cells, Therapy, Alzheimer's disease (AD)

1. Introduction

According to the World Alzheimer Report 2019, more than 50 million people worldwide suffer from dementia. It is expected to grow to 152 million by 2050. The current cost of treating dementia is $1 trillion a year, and that cost is expected to double by 2030. There are more than 200 subtypes of dementia, of which 50 to 60 percent are caused by Alzheimer's disease (AD). The concept of the disease was proposed by Alois Alzheimer in 1907. It was later recognized as the most common neurodegenerative disease. Although decades have passed since the discovery of the pathological mechanism of Alzheimer's disease, we still do not know what causes the disease. It is well known that Alzheimer's disease is a sporadic, age-related disease, with only a small proportion caused by genetic factors. The disease is characterized by a progressive decline in cognitive function. Clinically, these patients present with short-term memory impairments that interfere with activities of daily living, followed by impairments in other cognitive areas such as language, logical understanding, orientation, executive function, judgment, behavior, and finally motor impairments [1]. The pathological features of AD include: Senile Plaques (SP) formed by the deposition of amyloid β-protein (Aβ) outside neurocyte; The abnormal phosphorylation of intracellular Tau protein results in the neurofibrillary tangles (NFTs); Synaptic loss, neuroinflammation, neurocyte apoptosis in the neocortex and hippocampus of the brain. The pathological manifestations were brain atrophy [2–5]. In this review, we believe that the most effective strategies should target the biological feature which is most associated with symptoms, the loss of synapses, to treat the disease. Specifically, we focus on recent advances in cell-based therapies that aim at repopulation or regeneration of degenerating neuronal networks in AD [6].

2. Alzheimer's Disease's neuropathology

As mentioned in the background, we group the pathological changes of Alzheimer's disease into two types, which provide evidence of the disease's occurrence and progression: (1) Positive lesions. The main findings include SP caused by Aβ deposition and NFTs caused by abnormal phosphorylation of intracellular Tau protein. Otherwise, dystrophic neurites, neuropil threads and various other sediments found in the brains of patients with AD also falls into this category. (2) Negative lesions, which can also called loss type lesions. The main clinical manifestation is brain atrophy due to loss of synapses. At the same time, other factors, including neuroinflammation, oxidative stress, and damage to cholinergic neurons, are all important factors leading to the occurrence of neurodegenerative diseases.

2.1 Senile plaques (SP)

The SP are extracellular deposits of Aβ with different morphological forms, including neuritic, diffuse, dense-cored, or classic and compact type plaques [5].

The formation of Aβ is from the amyloidogenic cleavage of human amyloid precursor protein (APP) [7]. The anomalous processing of APP by β-secretases and γ-secretases leads to production of $A\beta_{40}$ and $A\beta_{42}$ monomers, which further oligomerize and aggregate into SP [8, 9]. Although soluble $A\beta_{40}$ is much more abundant than soluble $A\beta_{42}$, $A\beta_{42}$ exhibits a higher propensity for aggregation, due to hydrophobicity within its two terminal residues. Indeed, $A\beta_{42}$ is the main component of amyloid plaques and is shown to be neurotoxic [10]. Recent neuroimaging and neuropathology researches reveal that Aβ sedimentation is mainly related to cognitive disorder of the old, and it is not very relevant with other clinical features [11].

2.2 Neurofibrillary tangles (NFTs)

Tau protein is mainly distributed in neurons. Repeated Pro-Gly-Gly-Gly fragments help it bind to tubulin and maintain the structural stability of microtubules. The presence of Tau protein contributes to the maintenance of cytoskeleton and the integrity of axon transport [12]. NFTs are filamentous structures filled in the cytoplasm of neurons -- paired helical fibers (PHF). The reason of Tau hyperphosphorylation is the increased protein kinase activity. Protein kinase activity such as glycogen synthase kinase 3β (GSK-3β) activity can be decreased to reduce phosphorylation. Meanwhile, decreased phosphatase activity is also the reason of hyperphosphorylation. In addition, the lack of glucose in the brain can make Tau hyperphosphorylated by mediating the signal pathway of p38 mitogen-activated protein kinase (MAPK). Increasing the level of glucose in the brain may provide a new idea for treating AD, by using a pharmacological model of glucose deprivation and investigated its effect on Tau phosphorylation, synaptic function and cognition in a relevant transgenic mouse model of tauopathy, the h-Tau mouse [13]. It has been shown that phosphorylation of Tau protein at the early stage of AD inhibits Aβ toxicity, being that Tau phosphorylation-mediated by p38 MAPK can antagonize the postsynaptic excitation toxicity caused by Aβ [14, 15].

2.3 Synaptic loss

Soluble Aβ collaborate with pTau to induce synapse loss and cognitive impairment in AD [16]. Metabolism of Aβ and Tau proteins is crucially influenced by autophagy. Autophagy is a lysosome-dependent, homeostatic process, in which organelles and proteins are degraded and recycled into energy [17]. Neuroplasticity is an ongoing process that responds to the activity, injury, and death of neurons, including the regulation of the structure and function of axons, dendrites and synapses [18]. Overdeposition of Aβ and abnormal phosphorylation of Tau both lead to decreased neuroplasticity, which is manifested in a series of clinical symptoms caused by synaptic loss in AD [12, 19]. Aβ and Tau both trigger mitochondrial alterations. Some evidence suggests that mitochondrial perturbation acts as a key factor that is involved in synaptic failure and degeneration in AD [20]. Synaptic plasticity and long-term potentiation (LTP) are all about N-methyl-D-aspartate receptor (NMDAR). Aβ oligomer facilitates astrocytes (AS) to release glutamate by a7nAChR and activates NMDAR, making extracellular regulated protein kinases (ERK) signaling pathway to be suppressed and finally suppressing LTP, therefore the synaptic damages caused by NMDAR hyperactivation are the possible mechanisms of AD occurring [21].

2.4 Neuroinflammation

The AD pathophysiology entails chronic inflammation involving innate immune cells including microglia, astrocytes, and other peripheral blood cells. Inflammatory mediators such as cytokines and complements are also linked to AD pathogenesis [22, 23]. Activation of microglia can induce the production of inflammasomes, which in turn increase inflammatory cytokines, and may eventually result in Aβ deposition [24, 25]. Studies have shown that after being activated, astrocytes will release the corresponding cytokines, which can lead to the enhancement of neuronal toxicity, as well as a decreased outgrowth of neuronal processes and an overall decreased activity rate [26]. Recent studies have shown that there is a direct interaction between microglia and astrocytes. In the form that once microglia are activated, they can lead to activation of astrocytes, thus forming feed-forward loops that are harmful to the surrounding environment [26]. The mechanism showed that when being activated, microglia release IL-1α, TNFα and C1q and astrocytes become activated. Microglia and astrocytes are major modulators of inflammation in the brain, and they are also the major sources of apolipoprotein E (ApoE) in the brain. ApoE is a multifunctional protein with central roles in lipid metabolism. It transports lipids, including cholesterol, through the cerebrospinal fluid (CSF) and plasma [27, 28]. Earlier studies have shown that the presence of ApoE helps to inhibit glial activation of lipopolysaccharides in glial cell culture experiments, suggesting that ApoE may exert a protective anti-inflammatory effect [29, 30]. Moreover, the exacerbated proinflammatory state that occurs during this period of AD can trigger the hyperphosphorylation of Tau. Several of the kinases responsible for Tau phosphorylation are activated by proinflammatory mediators and have been shown to worsen Tau pathology [31].

2.5 Cholinergic neurons' injuries

The Acetylcholine (ACh) receptor (AChR) is a vital membrane protein on which ACh acts as a neurotransmitter. The cholinergic receptors are broadly categorized as muscarinic ACh receptors (mAChR) and nicotinic ACh receptors (nAChR) on the basis of their exogenous agonists [32]. ACh plays an important role in human memory function and is strongly associated with age-related dementia such as AD, in which hippocampal dependent learning dysfunction is prominent. Cholinergic neurons densely dominate the hippocampus and mediate the production of episodic and semantic memory [33]. In patients with AD, the synthesis, release and uptake of ACh in the hippocampus, neocortex and cerebrospinal fluid were decreased, the choline acetyltransferase (AChE) was significantly decreased, and the activity of acetylcholinesterase was decreased [34]. Clinically, the main method of drug treatment for AD is to improve the function of the brain's cholinergic system. Although inhibitors of acetylcholinesterase is a symptomatic relief treatment with marginal benefits, it is currently the most available clinical treatment which gives desperate AD patients a glimmer of hope [35].

3. Stem cell therapy for AD

There are some theoretical approaches to treat early AD. One is to target upregulation of resident neural stem cells (NSCs) niches within the adult brain. In fact, this regulation is to stimulate the development of adult hippocampal nerve, which has reached the purpose of compensating the degenerated nerve. Adult hippocampal neurogenesis may play a key role in learning and memory, so promoting this endogenous process may help improve amnesia in patients with early AD. Another approach is to up-regulate growth factors that are known to modulate neurogenesis integrally, either through drug therapy or gene therapy or, as we describe in this paper, through stem cell therapy. This type of growth factor includes brain-derived neurotrophic factor (BDNF) [36, 37], insulin growth factor-1 (IGF-1) [38], nerve growth factor (NGF) [39–42], vascular endothelial growth factor (VEGF) [43, 44] and so on. Stem cell therapy aims to rescue cognitive function by introducing exogenous stem cells to restore degenerated neural networks. These stem cells can be used as cell delivery systems through the natural or induced production of neuroprotective growth factors utilizing the paracrine "bystander" mechanism. Alternatively, therapeutic recovery may occur through the differentiation and involvement of stem cells in refilling degenerated neuronal circuits. It's a finely balanced, complex, multi-step process.

Some of stem cells are now in clinical use, such as embryonic stem cells (ESCs) derived from the inner cell mass of preimplantation embryos and induced pluripotent stem cells (iPSCs) derived from the epiblast layer of implanted embryos [45, 46]. Mesenchymal stem cells (MSCs) can promote tissue repair through the secretion of extracellular vesicles that carry a variety of cytokines, growth factors and microRNAs (miRNAs) [47]. Adipose tissue-derived stem cells (ADSCs) are a replacement therapy for MSCs, with the similar mechanism which secretion extracellular vesicles (EVs) to multiple proteins possessing neuroprotective and neurogenesis activities [48]. NSCs participate extensively in mammalian brain homeostasis and repair and exhibit pleiotropic intrinsic properties which makes them a good method for the treatment of AD [49].

3.1 ESCs

ESCs are cells isolated from early embryos or primitive gonads. It has the characteristics of infinite proliferation, self-renewal and multidirectional differentiation *in vitro* culture. Both *in vitro* and *in vivo*, ESCs can be induced to differentiate into almost all cell types in the body, so they can be used to improve the recovery of neurodegenerative diseases (such as AD). Therefore ESCs have a broad application prospect in autologous stem cell therapy [50, 51]. Thymic epithelial progenitor cells derived from mouse ESCs with deleted amyloid precursor protein gene have been proved to have the ability to alleviate AD symptoms [52]. Early human embryonic stem cells (hESCs)-derived neural populations consist of various embryonic neural progenitors (ENPs) with broad neural developmental propensity. The hESC-ENP-enriched neural transcription factors (TFs) can directly transform human cells into ENP phenotypes. Induced ENPs (IENPs) and their derivatives summarize the signature pathological characteristics of AD and hold promise for future strategies for disease modeling and clinical intervention [53].

Although ESCs are good candidates for AD cell therapy, they may bring some ethical and practical problems. Even if we overcome the problem of immune rejection, there have been reports of teratomas resulting from transplanted ESCs [54].

3.2 NSCs

NSCs have the ability to differentiate into neuronal astrocytes and oligodendrocytes, which are self-renewing and sufficient to provide a large number of brain tissue cells [55, 56]. In the past, it was thought that NSCs lost their ability to regenerate during the prenatal period or several months after birth. However, some recent studies have shown that NSCs also exist in adult brain tissues, mainly located in the subventricular zone (SVZ) and hippocampus dentate gyrus (DG) [57–60]. Due to their multidirectional differentiation and self-renewal, NSCs play an important role in maintaining brain homeostasis, promoting normal nerve development and repairing damaged nerves, which provides a possible choice for stem cell therapy for AD [49, 61]. A large number of studies have shown that the gradual accumulation of Aβ leading to the loss of synapses related to cognitive deficits is an important mechanism of AD [62]. In the hippocampus of AD mice after NSCs transplantation, the level of Synaptophysin (SYP), postsynaptic density protein 95 (PSD-95) and microtubule-associated protein (MAP-2) were significantly increased, which are important protein markers related to synaptic plasticity and play an important role in synaptic plasticity and stability, indicating improved learning and memory ability in AD mice [63–65]. Damage of cholinergic neurons in the basal forebrain is another important feature of AD [66]. Reduced cholinergic function due to cholinergic neuron injury may results in learning and memory impairments [67]. Transplantation of NSCs into the basal forebrain will increase the level of choline acetyltransferase (ChAT) protein, restoring the damaged neurons and improving the learning and memory ability [68, 69].

Recent studies have demonstrated the mechanism of NSCs transplantation to improve cognitive function, which is replacing damaged neurons with the differentiation of transplanted NSCs and enhancing synaptic density by releasing neurotrophic factors [61, 70, 71]. Neurotrophic factors have been shown to improve cognitive impairment [72, 73]. Although NSCs transplantation has great potential to be an excellent choice of cell therapy for AD in the future, there are many problems in its application: (1) The attribution that supports the differentiation of NSCs into a specific cell type is not clear. (2) Although NSCs transplantation can salvage synaptic damage and participate in the interaction of endogenous neuronal circuit function, there is no accurate answer to the duration of this effect. (3) The localization of the transplanted area and the viability of the transplanted cells are only the initial challenges of NSCs therapy, and subsequent interactions with cells in the host environment are also important. In some studies, NSCs after transplantation is difficult to trace, and in the cases where NSCs can be traced, the number of activated cells is also difficult to quantify [74]. (4) Many studies have identified transplanted NSCs have potential risk of developing brain tumors, such as glioblastoma [75, 76]. (5) Extrinsic NSCs transplantation also involves ethical issues. Direct isolation of NSCs from the primary tissue is dangerous. Non-patient-specific NSCs are more likely to result in immune rejection [54, 77].

3.3 MSCs

MSCs are pluripotent stem cells, which have all the common features of stem cells, namely self-renewal and multidirectional differentiation. As major stem cells that have undergone extensive clinical trials, MSCs bring hope for the treatment of a variety of diseases [78]. MSCs come from a wide range of sources. The most common ones are bone marrow mesenchymal stem cells (BMSCs), adipose-derived stem cells (ADSCs), umbilical cord derived mesenchymal stem cells (UC-MSCs), etc. Their biological characteristics are also different [79].

MSCs have the ability of immune regulation, neuroprotection and regeneration. The main mechanisms of MSCs in the treatment of AD are as follows [80]: (1) Secrete growth factors: MSCs secrete a variety of pro-cytokines that may play a beneficial role in AD [81]. (2) Secrete exosomes: Exosomes refer to extracellular vesicles, which are biocompatible nanoparticles with lipid membranes. These vesicles can transmit messages across biological barriers. Studies have shown that intercellular exchange of miRNA and proteins through EVs can reduce neuroinflammation, promote neurogenesis and angiogenesis, save learning disabilities and improve functional recovery [82, 83]. (3) Reduce neuroinflammation by regulating autophagy: MSCs can affect the autophagy of immune cells involved in injury-induced inflammation, thereby reducing their survival, proliferation and function, and facilitating the regression of inflammation. In addition, MSCs can affect the autophagy of endogenous adult or progenitor cells, promote their survival, proliferation and differentiation, and support the recovery of functional tissues [84]. In addition, foreign proteins conveyed by MSCs can regulate microglia function and enhance neurogenesis, so as to alleviate early memory deficits in AD [85]. Transplantation of MSCs carrying CX3CL1 (a multifunctional inflammatory chemokine with a single receptor CX3CR1) [86] and Wnt3a (CX3CL1-Wnt3a-MSC) can regulate phosphoinositide 3-kinase/activated protein kinase B (PI3K/AKT) signaling to inhibit the activity of glycogen synthase kinase 3 beta (GSK3β), improving the neurobehavioral function of mice by transplanting microglia with neurotoxicity and promoting hippocampal neurogenesis.

Reports have shown that EVs secreted by adipocytes derived from ADSCs may treat AD by alleviating neuronal damage, promoting neurogenesis and reducing the increase of neuronal apoptosis [48, 87]. EVs secreted by BMSCs can reach astrocytes to promote synaptic development and improve cognitive impairment [88, 89]. Hepatocyte growth factor (HGF), a core functional factor secreted by UC-MSCs, plays a key role in regulating the recovery of damaged nerve cells [90]. MSCs derived from ESCs have a better effect than BMSCs in the treatment of AD [91].

Modified MSCs pretreated with different conditions or reagents can significantly enhance the therapeutic effect of AD and improve cognitive impairment, such as cytokine pretreated MSCs [92], hypoxia preconditioned MSCs (PCMSCs) [93], MSCs modified by lin28B [94], MSCs prepared by cerebrospinal fluid of AD patients [95], ADSCs pretreated by melatonin (MT) [96], UC-MSCs combined with resveratrol [97].

3.4 iPSCs

Using defined reprogramming factors to reprogram fully differentiated somatic cells into iPSCs has become a novel strategy to produce pluripotent cells derived from patients that enable autologous transplantation [98]. The apolipoprotein E4 (ApoE4) variant is the single greatest genetic risk factor for sporadic Alzheimer's disease (sAD) [27–30]. sAD iPSCs convert ApoE4 to ApoE3 in brain cell types. This conversion can reduce many AD-related diseases [99]. The generation of neural precursors from iPSCs has also been extensively studied. In the production of astrocytes, the mutation in presenilin1 (PSEN1) increased Aβ production and oxidative stress. At the same time, it also altered cytokine release and Ca^{2+} homeostasis. These changes reducing neuronal support function in PSEN1 astrocytes [100, 101]. EVs of either

50–200 nm in size (called exosomes) or 200 nm^{-1} μm in size (called micro-vesicles) are membrane-bounded vesicles. They can carry RNAs, proteins, and other metabolites. They are secreted from all cell types and present in biological fluids such as serum and plasma [50, 102]. Human iPSCs can be cultured infinitely under a chemically defined medium. The properties and functions of exosomes and micro-vesicles (called EMVs) from human iPSCs are different with the ones secreted by human MSCs. Purified EVs produced by both stem cell types have similar sizes, but human iPSCs produced 16-fold more EVs than MSCs [103]. Neurons from patients with early-onset familial Alzheimer's disease (fAD) and patients with late-onset sAD showed increased phosphorylation of Tau protein at all investigated phosphorylation sites. Relative to the control neurons, neurons derived from patients with fAD and patients with sAD exhibited higher levels of extracellular amyloid-β 1–40 ($Aβ_{1–40}$) and amyloid-β 1–42 ($Aβ_{1–42}$) [104–106]. Using iPSCs-derived neurons to recapitulate AD pathology *in vitro* has significant applications in the study of pathogenesis and screening for potential therapeutic drugs. They are now the subject of extensive study *in vitro* [107]. Studies have also shown that EVs from iPSCs can play an important role in heart repair [108].

3.5 Clinical trials and results in humans

Due to the inconsistent results of various preclinical studies, stem cell therapies other than MSCs are still difficult to be applied clinically. Some articles specifically showed us the application of MSCs-based stem cell therapy in human clinical trials [6, 80, 109, 110]. In recent years, more studies have been conducted on rodents. The effects of MSCs on AD pathology and cognitive mouse models may be mediated by the regulation of neuroinflammation [111, 112]. In recent years, clinical trials using mesenchymal stem cells have been conducted around the world. A completed clinical trial in the United States (Trial identifier: NCT03117738) investigated the safety and efficacy of autologous ADSCs. At the same time, a team studied the efficacy of UC-MSCs (Trial identifier: NCT01297218). Compared with cholinergic drugs that only improve symptoms, UC-MSCs are immunologically stable and not-toxic, and have better therapeutic effect on AD. UC-MSCS remain a common cell choice, although there are key differences in cell number, dose quantity, and dose schedule (Trial identifier: NCT03172117). Two separate trials, both currently undergoing recruitment, will utilize alternative MSC sources. One studies human MSCs (Trial identifier: NCT02833792) and evaluates its safety and efficacy. The other utilizes the exosomes derived from allogenic adipose mesenchymal stem cells (MSCs-Exos) (Trial identifier: NCT04388982) to treat patients with mild to moderate dementia due to AD. While many of these trials employ an intravenous infusion administration route, one trial (Trial identifier: NCT03724136) administered BMSCs to the nasal mucosa topically, to investigate whether there was an improvement in efficacy in combination with intravenous injection.

4. Future directions

Numerous preclinical studies have revealed the different mechanisms of various stem cells and demonstrated the great potential of stem cells to treat AD. However, the biggest problem in this area of research is that it is difficult to translate animal studies into human trials. In fact, researchers have used nearly a hundred methods to effectively treat AD in transgenic mouse models. Disappointingly, almost every approach has failed in human clinical trials or has never even been tested in humans. Clearly, rodent models and their pathological assumptions are insufficient to predict clinical outcomes in humans. Therefore, the establishment of more accurate models is needed for cell therapy of AD. Since the goal of truly simulating the pathological progress of AD in human body has been achieved, more experiments on cell therapy need to be carried out.

At the same time, key questions remain to be addressed, including the safety of treatment, optimal cell source and delivery system. While cell therapies may not be able to fully compensate for the loss of extensive synapses, they can help to temporarily improve existing depleted circuits enough to improve cognitive function, restore basic daily living functions, and improve quality of life. For us, stem cell therapy for AD still has a long way to go.

5. Conclusion

AD is a neurodegenerative disease, which is characterized by excessive deposition of Aβ and abnormal phosphorylation of Tau protein and synaptic loss. Studies and clinical trials in recent years are also based on these basic mechanisms. Although the role of stem cell therapy in AD is not fully understood, many preclinical studies have provided a number of promising results. However, human clinical trials are still in their infancy, and most current research is still centered on animal experiments. But it also shows the broad prospects of stem cell therapy for the AD. A large number of preclinical studies have demonstrated the theoretical basis, and new studies are continuing to reveal the underlying mechanisms. Among many stem cells, MSCs-based therapies are widely accepted and have met certain clinical trial standards. The vast majority of cell therapies for AD have been conducted on rodents, and we must be aware of a wide range of physiological differences between humans and rodents. We need to understand the mechanism of treatment through animal experiments and establish the correct translation model for human application.

Author details

Kun Jiang, Yongqi Zhu and Lei Zhang*
Department of Human Anatomy, Institute of Neurobiology, Jiangsu Key Laboratory of Neuroregeneration, Medical School, Nantong University, Nantong, Jiangsu Province, China

*Address all correspondence to: zhanglei@ntu.edu.cn

References

[1] Vasic, V.; Barth, K.; Schmidt, M. H. H. Neurodegeneration and Neuro-Regeneration-Alzheimer's Disease and Stem Cell Therapy. Int J Mol Sci **2019**, *20*.

[2] Zhang, H.; Zheng, Y. [beta Amyloid Hypothesis in Alzheimer's Disease: Pathogenesis, Prevention, and Management]. Zhongguo Yi Xue Ke Xue Yuan Xue Bao **2019**, *41*, 702-708.

[3] Soria Lopez, J. A.; Gonzalez, H. M.; Leger, G. C. Alzheimer's disease. Handb Clin Neurol **2019**, *167*, 231-255.

[4] Esquerda-Canals, G.; Montoliu-Gaya, L.; Guell-Bosch, J.; Villegas, S. Mouse Models of Alzheimer's Disease. J Alzheimers Dis **2017**, *57*, 1171-1183.

[5] Breijyeh, Z.; Karaman, R. Comprehensive Review on Alzheimer's Disease: Causes and Treatment. Molecules **2020**, *25*.

[6] Duncan, T.; Valenzuela, M. Alzheimer's disease, dementia, and stem cell therapy. Stem Cell Res Ther **2017**, *8*, 111.

[7] Reiss, A. B.; Arain, H. A.; Stecker, M. M.; Siegart, N. M.; Kasselman, L. J. Amyloid toxicity in Alzheimer's disease. Rev Neurosci **2018**, *29*, 613-627.

[8] Tiwari, S.; Atluri, V.; Kaushik, A.; Yndart, A.; Nair, M. Alzheimer's disease: pathogenesis, diagnostics, and therapeutics. Int J Nanomedicine **2019**, *14*, 5541-5554.

[9] Gallardo, G.; Holtzman, D. M. Amyloid-beta and Tau at the Crossroads of Alzheimer's Disease. Adv Exp Med Biol **2019**, *1184*, 187-203.

[10] Guo, T.; Zhang, D.; Zeng, Y.; Huang, T. Y.; Xu, H.; Zhao, Y. Molecular and cellular mechanisms underlying the pathogenesis of Alzheimer's disease. Mol Neurodegener **2020**, *15*, 40.

[11] Shimada, H.; Kitamura, S.; Shinotoh, H.; Endo, H.; Niwa, F.; Hirano, S.; Kimura, Y.; Zhang, M. R.; Kuwabara, S.; Suhara, T.; Higuchi, M. Association between Abeta and tau accumulations and their influence on clinical features in aging and Alzheimer's disease spectrum brains: A [(11)C]PBB3-PET study. Alzheimers Dement (Amst) **2017**, *6*, 11-20.

[12] Chen, Y. G. Research Progress in the Pathogenesis of Alzheimer's Disease. Chin Med J (Engl) **2018**, *131*, 1618-1624.

[13] Lauretti, E.; Li, J. G.; Di Meco, A.; Pratico, D. Glucose deficit triggers tau pathology and synaptic dysfunction in a tauopathy mouse model. Transl Psychiatry **2017**, *7*, e1020.

[14] Lee, S. H.; Le Pichon, C. E.; Adolfsson, O.; Gafner, V.; Pihlgren, M.; Lin, H.; Solanoy, H.; Brendza, R.; Ngu, H.; Foreman, O.; Chan, R.; Ernst, J. A.; DiCara, D.; Hotzel, I.; Srinivasan, K.; Hansen, D. V.; Atwal, J.; Lu, Y.; Bumbaca, D.; Pfeifer, A.; Watts, R. J.; Muhs, A.; Scearce-Levie, K.; Ayalon, G. Antibody-Mediated Targeting of Tau In Vivo Does Not Require Effector Function and Microglial Engagement. Cell Rep **2016**, *16*, 1690-1700.

[15] Ittner, A.; Chua, S. W.; Bertz, J.; Volkerling, A.; van der Hoven, J.; Gladbach, A.; Przybyla, M.; Bi, M.; van Hummel, A.; Stevens, C. H.; Ippati, S.; Suh, L. S.; Macmillan, A.; Sutherland, G.; Kril, J. J.; Silva, A. P.; Mackay, J. P.; Poljak, A.; Delerue, F.; Ke, Y. D.; Ittner, L. M. Site-specific phosphorylation of tau inhibits amyloid-beta toxicity in Alzheimer's mice. Science **2016**, *354*, 904-908.

[16] Jeong, S. Molecular and Cellular Basis of Neurodegeneration in Alzheimer's Disease. Mol Cells **2017**, *40*, 613-620.

[17] Uddin, M. S.; Stachowiak, A.; Mamun, A. A.; Tzvetkov, N. T.; Takeda, S.; Atanasov, A. G.; Bergantin, L. B.; Abdel-Daim, M. M.; Stankiewicz, A. M. Autophagy and Alzheimer's Disease: From Molecular Mechanisms to Therapeutic Implications. Front Aging Neurosci **2018**, *10*, 04.

[18] Skaper, S. D.; Facci, L.; Zusso, M.; Giusti, P. Synaptic Plasticity, Dementia and Alzheimer Disease. CNS Neurol Disord Drug Targets **2017**, *16*, 220-233.

[19] Colom-Cadena, M.; Spires-Jones, T.; Zetterberg, H.; Blennow, K.; Caggiano, A.; DeKosky, S. T.; Fillit, H.; Harrison, J. E.; Schneider, L. S.; Scheltens, P.; de Haan, W.; Grundman, M.; van Dyck, C. H.; Izzo, N. J.; Catalano, S. M.; Synaptic Health Endpoints Working, G. The clinical promise of biomarkers of synapse damage or loss in Alzheimer's disease. *Alzheimers Res Ther* **2020**, *12*, 21.

[20] Cai, Q.; Tammineni, P. Mitochondrial Aspects of Synaptic Dysfunction in Alzheimer's Disease. J Alzheimers Dis **2017**, *57*, 1087-1103.

[21] Talantova, M.; Sanz-Blasco, S.; Zhang, X.; Xia, P.; Akhtar, M. W.; Okamoto, S.; Dziewczapolski, G.; Nakamura, T.; Cao, G.; Pratt, A. E.; Kang, Y. J.; Tu, S.; Molokanova, E.; McKercher, S. R.; Hires, S. A.; Sason, H.; Stouffer, D. G.; Buczynski, M. W.; Solomon, J. P.; Michael, S.; Powers, E. T.; Kelly, J. W.; Roberts, A.; Tong, G.; Fang-Newmeyer, T.; Parker, J.; Holland, E. A.; Zhang, D.; Nakanishi, N.; Chen, H. S.; Wolosker, H.; Wang, Y.; Parsons, L. H.; Ambasudhan, R.; Masliah, E.; Heinemann, S. F.; Pina-Crespo, J. C.; Lipton, S. A. Abeta induces astrocytic glutamate release, extrasynaptic NMDA receptor activation, and synaptic loss. *Proc Natl Acad Sci U S A* **2013**, *110*, E2518-2527.

[22] Park, J. C.; Han, S. H.; Mook-Jung, I. Peripheral inflammatory biomarkers in Alzheimer's disease: a brief review. BMB Rep **2020**, *53*, 10-19.

[23] Regen, F.; Hellmann-Regen, J.; Costantini, E.; Reale, M. Neuroinflammation and Alzheimer's Disease: Implications for Microglial Activation. Curr Alzheimer Res **2017**, *14*, 1140-1148.

[24] Kloske, C. M.; Wilcock, D. M. The Important Interface Between Apolipoprotein E and Neuroinflammation in Alzheimer's Disease. Front Immunol **2020**, *11*, 754.

[25] Houtman, J.; Freitag, K.; Gimber, N.; Schmoranzer, J.; Heppner, F. L.; Jendrach, M. Beclin1-driven autophagy modulates the inflammatory response of microglia via NLRP3. EMBO J **2019**, *38*.

[26] Liddelow, S. A.; Guttenplan, K. A.; Clarke, L. E.; Bennett, F. C.; Bohlen, C. J.; Schirmer, L.; Bennett, M. L.; Munch, A. E.; Chung, W. S.; Peterson, T. C.; Wilton, D. K.; Frouin, A.; Napier, B. A.; Panicker, N.; Kumar, M.; Buckwalter, M. S.; Rowitch, D. H.; Dawson, V. L.; Dawson, T. M.; Stevens, B.; Barres, B. A. Neurotoxic reactive astrocytes are induced by activated microglia. Nature **2017**, *541*, 481-487.

[27] Yin, Y.; Wang, Z. ApoE and Neurodegenerative Diseases in Aging. Adv Exp Med Biol **2018**, *1086*, 77-92.

[28] Yamazaki, Y.; Zhao, N.; Caulfield, T. R.; Liu, C. C.; Bu, G. Apolipoprotein E and Alzheimer disease: pathobiology and targeting strategies. Nat Rev Neurol **2019**, *15*, 501-518.

[29] Serrano-Pozo, A.; Das, S.; Hyman, B. T. APOE and Alzheimer's disease: advances in genetics, pathophysiology, and therapeutic approaches. Lancet Neurol **2021**, *20*, 68-80.

[30] Zhao, N.; Liu, C. C.; Qiao, W.; Bu, G. Apolipoprotein E, Receptors, and Modulation of Alzheimer's Disease. Biol Psychiatry **2018**, *83*, 347-357.

[31] Huat, T. J.; Camats-Perna, J.; Newcombe, E. A.; Valmas, N.; Kitazawa, M.; Medeiros, R. Metal Toxicity Links to Alzheimer's Disease and Neuroinflammation. J Mol Biol **2019**, *431*, 1843-1868.

[32] Ahmed, T.; Zahid, S.; Mahboob, A.; Farhat, S. M. Cholinergic System and Post-translational Modifications: An Insight on the Role in Alzheimer's Disease. Curr Neuropharmacol **2017**, *15*, 480-494.

[33] Haam, J.; Yakel, J. L. Cholinergic modulation of the hippocampal region and memory function. *J Neurochem* **2017**, *142 Suppl 2*, 111-121.

[34] Pepeu, G.; Grazia Giovannini, M. The fate of the brain cholinergic neurons in neurodegenerative diseases. Brain Res **2017**, *1670*, 173-184.

[35] Du, X.; Wang, X.; Geng, M. Alzheimer's disease hypothesis and related therapies. Transl Neurodegener **2018**, *7*, 2.

[36] Hu, W.; Feng, Z.; Xu, J.; Jiang, Z.; Feng, M. Brain-derived neurotrophic factor modified human umbilical cord mesenchymal stem cells-derived cholinergic-like neurons improve spatial learning and memory ability in Alzheimer's disease rats. Brain Res **2019**, *1710*, 61-73.

[37] Pramanik, S.; Sulistio, Y. A.; Heese, K. Neurotrophin Signaling and Stem Cells-Implications for Neurodegenerative Diseases and Stem Cell Therapy. Mol Neurobiol **2017**, *54*, 7401-7459.

[38] Skop, N. B.; Singh, S.; Antikainen, H.; Saqcena, C.; Calderon, F.; Rothbard, D. E.; Cho, C. H.; Gandhi, C. D.; Levison, S. W.; Dobrowolski, R. Subacute Transplantation of Native and Genetically Engineered Neural Progenitors Seeded on Microsphere Scaffolds Promote Repair and Functional Recovery After Traumatic Brain Injury. ASN Neuro **2019**, *11*, 1759091419830186.

[39] Karimipour, M.; Rahbarghazi, R.; Tayefi, H.; Shimia, M.; Ghanadian, M.; Mahmoudi, J.; Bagheri, H. S. Quercetin promotes learning and memory performance concomitantly with neural stem/progenitor cell proliferation and neurogenesis in the adult rat dentate gyrus. Int J Dev Neurosci **2019**, *74*, 18-26.

[40] Wen, C.; Huang, C.; Yang, M.; Fan, C.; Li, Q.; Zhao, J.; Gan, D.; Li, A.; Zhu, L.; Lu, D. The Secretion from Bone Marrow Mesenchymal Stem Cells Pretreated with Berberine Rescues Neurons with Oxidative Damage Through Activation of the Keap1-Nrf2-HO-1 Signaling Pathway. Neurotox Res **2020**, *38*, 59-73.

[41] Vymetalova, L.; Kucirkova, T.; Knopfova, L.; Pospisilova, V.; Kasko, T.; Lejdarova, H.; Makaturova, E.; Kuglik, P.; Oralova, V.; Matalova, E.; Benes, P.; Koristek, Z.; Forostyak, S. Large-Scale Automated Hollow-Fiber Bioreactor Expansion of Umbilical Cord-Derived Human Mesenchymal Stromal Cells for Neurological Disorders. Neurochem Res **2020**, *45*, 204-214.

[42] Morelli, A.; Sarchielli, E.; Guarnieri, G.; Coppi, E.; Pantano, D.; Comeglio, P.; Nardiello, P.; Pugliese, A. M.; Ballerini, L.; Matucci, R.; Ambrosini, S.; Castronovo, G.; Valente, R.; Mazzanti, B.; Bucciantini, S.; Maggi, M.; Casamenti, F.; Gallina, P.; Vannelli, G. B. Young Human Cholinergic Neurons Respond to Physiological Regulators and Improve Cognitive Symptoms in an Animal Model of Alzheimer's Disease. Front Cell Neurosci **2017**, *11*, 339.

[43] Jin, K.; Zhu, Y.; Sun, Y.; Mao, X. O.; Xie, L.; Greenberg, D. A. Vascular endothelial growth factor (VEGF) stimulates neurogenesis in vitro and in vivo. Proc Natl Acad Sci U S A **2002**, *99*, 11946-11950.

[44] Li, B.; Gao, Y.; Zhang, W.; Xu, J. R. Regulation and effects of neurotrophic factors after neural stem cell transplantation in a transgenic mouse model of Alzheimer disease. J Neurosci Res **2018**, *96*, 828-840.

[45] Bloor, A. J. C.; Patel, A.; Griffin, J. E.; Gilleece, M. H.; Radia, R.; Yeung, D. T.; Drier, D.; Larson, L. S.; Uenishi, G. I.; Hei, D.; Kelly, K.; Slukvin, I.; Rasko, J. E. J. Production, safety and efficacy of iPSC-derived mesenchymal stromal cells in acute steroid-resistant graft versus host disease: a phase I, multicenter, open-label, dose-escalation study. Nat Med **2020**, *26*, 1720-1725.

[46] Morishima, Y.; Morishima, S.; Murata, M.; Arima, N.; Uchida, N.; Sugio, Y.; Takahashi, S.; Matsuhashi, Y.; Onizuka, M.; Eto, T.; Nagafuji, K.; Onishi, Y.; Inoue, M.; Atsuta, Y.; Fukuda, T.; Ichinohe, T.; Kato, S.; Kanda, J. Impact of Homozygous Conserved Extended HLA Haplotype on Single Cord Blood Transplantation: Lessons for Induced Pluripotent Stem Cell Banking and Transplantation in Allogeneic Settings. Biol Blood Marrow Transplant **2020**, *26*, 132-138.

[47] Keshtkar, S.; Azarpira, N.; Ghahremani, M. H. Mesenchymal stem cell-derived extracellular vesicles: novel frontiers in regenerative medicine. Stem Cell Res Ther **2018**, *9*, 63.

[48] Ma, X.; Huang, M.; Zheng, M.; Dai, C.; Song, Q.; Zhang, Q.; Li, Q.; Gu, X.; Chen, H.; Jiang, G.; Yu, Y.; Liu, X.; Li, S.; Wang, G.; Chen, H.; Lu, L.; Gao, X. ADSCs-derived extracellular vesicles alleviate neuronal damage, promote neurogenesis and rescue memory loss in mice with Alzheimer's disease. J Control Release **2020**, *327*, 688-702.

[49] Boese, A. C.; Hamblin, M. H.; Lee, J. P. Neural stem cell therapy for neurovascular injury in Alzheimer's disease. Exp Neurol **2020**, *324*, 113112.

[50] Kolagar, T. A.; Farzaneh, M.; Nikkar, N.; Khoshnam, S. E. Human Pluripotent Stem Cells in Neurodegenerative Diseases: Potentials, Advances and Limitations. Curr Stem Cell Res Ther **2020**, *15*, 102-110.

[51] De Filippis, L.; Zalfa, C.; Ferrari, D. Neural Stem Cells and Human Induced Pluripotent Stem Cells to Model Rare CNS Diseases. CNS Neurol Disord Drug Targets **2017**, *16*, 915-926.

[52] Zhao, J.; Su, M.; Lin, Y.; Liu, H.; He, Z.; Lai, L. Administration of Amyloid Precursor Protein Gene Deleted Mouse ESC-Derived Thymic Epithelial Progenitors Attenuates Alzheimer's Pathology. Front Immunol **2020**, *11*, 1781.

[53] Hou, P. S.; Chuang, C. Y.; Yeh, C. H.; Chiang, W.; Liu, H. J.; Lin, T. N.; Kuo, H. C. Direct Conversion of Human Fibroblasts into Neural Progenitors Using Transcription Factors Enriched in Human ESC-Derived Neural Progenitors. Stem Cell Reports **2017**, *8*, 54-68.

[54] Sugaya, K.; Vaidya, M. Stem Cell Therapies for Neurodegenerative Diseases. Adv Exp Med Biol **2018**, *1056*, 61-84.

[55] Han, F.; Bi, J.; Qiao, L.; Arancio, O. Stem Cell Therapy for Alzheimer's Disease. Adv Exp Med Biol **2020**, *1266*, 39-55.

[56] Wong, R. S. Y.; Cheong, S. K. Therapeutic potentials of neural stem cells in Alzheimer's disease. Malays J Pathol **2020**, *42*, 157-170.

[57] Zhang, W.; Gu, G. J.; Zhang, Q.; Liu, J. H.; Zhang, B.; Guo, Y.; Wang, M. Y.; Gong, Q. Y.; Xu, J. R. NSCs promote hippocampal neurogenesis, metabolic changes and synaptogenesis in APP/PS1 transgenic mice. Hippocampus **2017**, *27*, 1250-1263.

[58] Bond, A. M.; Ming, G. L.; Song, H. Adult Mammalian Neural Stem Cells and Neurogenesis: Five Decades Later. Cell Stem Cell **2015**, *17*, 385-395.

[59] Guo, W.; Patzlaff, N. E.; Jobe, E. M.; Zhao, X. Isolation of multipotent neural stem or progenitor cells from both the dentate gyrus and subventricular zone of a single adult mouse. Nat Protoc **2012**, *7*, 2005-2012.

[60] Lee, J. P.; McKercher, S.; Muller, F. J.; Snyder, E. Y. Neural stem cell transplantation in mouse brain. *Curr Protoc Neurosci* **2008**, *Chapter 3*, Unit 3 10.

[61] Marsh, S. E.; Blurton-Jones, M. Neural stem cell therapy for neurodegenerative disorders: The role of neurotrophic support. Neurochem Int **2017**, *106*, 94-100.

[62] Shankar, G. M.; Li, S.; Mehta, T. H.; Garcia-Munoz, A.; Shepardson, N. E.; Smith, I.; Brett, F. M.; Farrell, M. A.; Rowan, M. J.; Lemere, C. A.; Regan, C. M.; Walsh, D. M.; Sabatini, B. L.; Selkoe, D. J. Amyloid-beta protein dimers isolated directly from Alzheimer's brains impair synaptic plasticity and memory. Nat Med **2008**, *14*, 837-842.

[63] Zhu, Q.; Zhang, N.; Hu, N.; Jiang, R.; Lu, H.; Xuan, A.; Long, D.; Chen, Y. Neural stem cell transplantation improves learning and memory by protecting cholinergic neurons and restoring synaptic impairment in an amyloid precursor protein/presenilin 1 transgenic mouse model of Alzheimer's disease. Mol Med Rep **2020**, *21*, 1172-1180.

[64] Schmitt, U.; Tanimoto, N.; Seeliger, M.; Schaeffel, F.; Leube, R. E. Detection of behavioral alterations and learning deficits in mice lacking synaptophysin. Neuroscience **2009**, *162*, 234-243.

[65] Tu, S.; Okamoto, S.; Lipton, S. A.; Xu, H. Oligomeric Abeta-induced synaptic dysfunction in Alzheimer's disease. Mol Neurodegener **2014**, *9*, 48.

[66] Hampel, H.; Mesulam, M. M.; Cuello, A. C.; Farlow, M. R.; Giacobini, E.; Grossberg, G. T.; Khachaturian, A. S.; Vergallo, A.; Cavedo, E.; Snyder, P. J.; Khachaturian, Z. S. The cholinergic system in the pathophysiology and treatment of Alzheimer's disease. Brain **2018**, *141*, 1917-1933.

[67] Solari, N.; Hangya, B. Cholinergic modulation of spatial learning, memory and navigation. Eur J Neurosci **2018**, *48*, 2199-2230.

[68] Park, D.; Choi, E. K.; Cho, T. H.; Joo, S. S.; Kim, Y. B. Human Neural Stem Cells Encoding ChAT Gene Restore Cognitive Function via Acetylcholine Synthesis, Abeta Elimination, and Neuroregeneration in APPswe/PS1dE9 Mice. Int J Mol Sci **2020**, *21*.

[69] Chen, Y.; Pan, C.; Xuan, A.; Xu, L.; Bao, G.; Liu, F.; Fang, J.; Long, D. Treatment Efficacy of NGF Nanoparticles Combining Neural Stem Cell Transplantation on Alzheimer's Disease Model Rats. Med Sci Monit **2015**, *21*, 3608-3615.

[70] Wu, C. C.; Lien, C. C.; Hou, W. H.; Chiang, P. M.; Tsai, K. J. Gain of BDNF Function in Engrafted Neural Stem Cells Promotes the Therapeutic Potential for Alzheimer's Disease. Sci Rep **2016**, *6*, 27358.

[71] Kim, D. H.; Lim, H.; Lee, D.; Choi, S. J.; Oh, W.; Yang, Y. S.; Oh, J. S.; Hwang, H. H.; Jeon, H. B. Thrombospondin-1 secreted by human umbilical cord blood-derived mesenchymal stem cells rescues neurons from synaptic dysfunction in Alzheimer's disease model. Sci Rep **2018**, *8*, 354.

[72] Di Carlo, P.; Punzi, G.; Ursini, G. Brain-derived neurotrophic factor and schizophrenia. Psychiatr Genet **2019**, *29*, 200-210.

[73] Lima Giacobbo, B.; Doorduin, J.; Klein, H. C.; Dierckx, R.; Bromberg, E.; de Vries, E. F. J. Brain-Derived Neurotrophic Factor in Brain Disorders: Focus on Neuroinflammation. Mol Neurobiol **2019**, *56*, 3295-3312.

[74] Zheng, Y.; Huang, J.; Zhu, T.; Li, R.; Wang, Z.; Ma, F.; Zhu, J. Stem Cell Tracking Technologies for Neurological Regenerative Medicine Purposes. Stem Cells Int **2017**, *2017*, 2934149.

[75] Gimple, R. C.; Bhargava, S.; Dixit, D.; Rich, J. N. Glioblastoma stem cells: lessons from the tumor hierarchy in a lethal cancer. Genes Dev **2019**, *33*, 591-609.

[76] Yelle, N.; Bakhshinyan, D.; Venugopal, C.; Singh, S. K. Introduction to Brain Tumor Stem Cells. Methods Mol Biol **2019**, *1869*, 1-9.

[77] Hayashi, Y.; Lin, H. T.; Lee, C. C.; Tsai, K. J. Effects of neural stem cell transplantation in Alzheimer's disease models. J Biomed Sci **2020**, *27*, 29.

[78] Wei, X.; Yang, X.; Han, Z. P.; Qu, F. F.; Shao, L.; Shi, Y. F. Mesenchymal stem cells: a new trend for cell therapy. Acta Pharmacol Sin **2013**, *34*, 747-754.

[79] Hass, R.; Kasper, C.; Bohm, S.; Jacobs, R. Different populations and sources of human mesenchymal stem cells (MSC): A comparison of adult and neonatal tissue-derived MSC. Cell Commun Signal **2011**, *9*, 12.

[80] Staff, N. P.; Jones, D. T.; Singer, W. Mesenchymal Stromal Cell Therapies for Neurodegenerative Diseases. Mayo Clin Proc **2019**, *94*, 892-905.

[81] Park, S. E.; Lee, N. K.; Na, D. L.; Chang, J. W. Optimal mesenchymal stem cell delivery routes to enhance neurogenesis for the treatment of Alzheimer's disease: optimal MSCs delivery routes for the treatment of AD. Histol Histopathol **2018**, *33*, 533-541.

[82] Reza-Zaldivar, E. E.; Hernandez-Sapiens, M. A.; Minjarez, B.; Gutierrez-Mercado, Y. K.; Marquez-Aguirre, A. L.; Canales-Aguirre, A. A. Potential Effects of MSC-Derived Exosomes in Neuroplasticity in Alzheimer's Disease. Front Cell Neurosci **2018**, *12*, 317.

[83] Izadpanah, M.; Dargahi, L.; Ai, J.; Asgari Taei, A.; Ebrahimi Barough, S.; Mowla, S. J.; TavoosiDana, G.; Farahmandfar, M. Extracellular Vesicles as a Neprilysin Delivery System Memory Improvement in Alzheimer's Disease. Iran J Pharm Res **2020**, *19*, 45-60.

[84] Ceccariglia, S.; Cargnoni, A.; Silini, A. R.; Parolini, O. Autophagy: a potential key contributor to the therapeutic action of mesenchymal stem cells. Autophagy **2020**, *16*, 28-37.

[85] Li, A.; Zhao, J.; Fan, C.; Zhu, L.; Huang, C.; Li, Q.; Gan, D.; Wen, C.; Chen, M.; Lu, D. Delivery of exogenous proteins by mesenchymal stem cells attenuates early memory deficits in a murine model of Alzheimer's disease. Neurobiol Aging **2020**, *86*, 81-91.

[86] Conroy, M. J.; Lysaght, J. CX3CL1 Signaling in the Tumor Microenvironment. Adv Exp Med Biol **2020**, *1231*, 1-12.

[87] Lee, M.; Ban, J. J.; Yang, S.; Im, W.; Kim, M. The exosome of adipose-derived stem cells reduces beta-amyloid pathology and apoptosis of neuronal cells derived from the transgenic mouse model of Alzheimer's disease. Brain Res **2018**, *1691*, 87-93.

[88] Nakano, M.; Kubota, K.; Kobayashi, E.; Chikenji, T. S.; Saito, Y.; Konari, N.; Fujimiya, M. Bone marrow-derived mesenchymal stem cells improve cognitive impairment in an Alzheimer's disease model by increasing the expression of microRNA-146a in hippocampus. Sci Rep **2020**, *10*, 10772.

[89] Elia, C. A.; Tamborini, M.; Rasile, M.; Desiato, G.; Marchetti, S.; Swuec, P.; Mazzitelli, S.; Clemente, F.; Anselmo, A.; Matteoli, M.; Malosio, M. L.; Coco, S. Intracerebral Injection of Extracellular Vesicles from Mesenchymal Stem Cells Exerts Reduced Abeta Plaque Burden in Early Stages of a Preclinical Model of Alzheimer's Disease. Cells **2019**, *8*.

[90] Jia, Y.; Cao, N.; Zhai, J.; Zeng, Q.; Zheng, P.; Su, R.; Liao, T.; Liu, J.; Pei, H.; Fan, Z.; Zhou, J.; Xi, J.; He, L.; Chen, L.; Nan, X.; Yue, W.; Pei, X. HGF Mediates Clinical-Grade Human Umbilical Cord-Derived Mesenchymal Stem Cells Improved Functional Recovery in a Senescence-Accelerated Mouse Model of Alzheimer's Disease. Adv Sci (Weinh) **2020**, *7*, 1903809.

[91] Kim, D. Y.; Choi, S. H.; Lee, J. S.; Kim, H. J.; Kim, H. N.; Lee, J. E.; Shin, J. Y.; Lee, P. H. Feasibility and Efficacy of Intra-Arterial Administration of Embryonic Stem Cell Derived-Mesenchymal Stem Cells in Animal Model of Alzheimer's Disease. J Alzheimers Dis **2020**, *76*, 1281-1296.

[92] Losurdo, M.; Pedrazzoli, M.; D'Agostino, C.; Elia, C. A.; Massenzio, F.; Lonati, E.; Mauri, M.; Rizzi, L.; Molteni, L.; Bresciani, E.; Dander, E.; D'Amico, G.; Bulbarelli, A.; Torsello, A.; Matteoli, M.; Buffelli, M.; Coco, S. Intranasal delivery of mesenchymal stem cell-derived extracellular vesicles exerts immunomodulatory and neuroprotective effects in a 3xTg model of Alzheimer's disease. Stem Cells Transl Med **2020**, *9*, 1068-1084.

[93] Cui, G. H.; Wu, J.; Mou, F. F.; Xie, W. H.; Wang, F. B.; Wang, Q. L.; Fang, J.; Xu, Y. W.; Dong, Y. R.; Liu, J. R.; Guo, H. D. Exosomes derived from hypoxia-preconditioned mesenchymal stromal cells ameliorate cognitive decline by rescuing synaptic dysfunction and regulating inflammatory responses in APP/PS1 mice. FASEB J **2018**, *32*, 654-668.

[94] Wu, K.; Zhang, R.; Lu, Y.; Wen, L.; Li, Y.; Duan, R.; Yao, Y.; Jia, Y. Lin28B regulates the fate of grafted mesenchymal stem cells and enhances their protective effects against Alzheimer's disease by upregulating IGF-2. J Cell Physiol **2019**, *234*, 21860-21876.

[95] Lee, J.; Kwon, S. J.; Kim, J. H.; Jang, H.; Lee, N. K.; Hwang, J. W.; Kim, J. H.; Chang, J. W.; Na, D. L. Cerebrospinal fluid from Alzheimer's disease patients as an optimal formulation for therapeutic application of mesenchymal stem cells in Alzheimer's disease. Sci Rep **2019**, *9*, 564.

[96] Nasiri, E.; Alizadeh, A.; Roushandeh, A. M.; Gazor, R.; Hashemi-Firouzi, N.; Golipoor, Z. Melatonin-pretreated adipose-derived mesenchymal stem cells efficeintly improved learning, memory, and cognition in an animal model of Alzheimer's disease. Metab Brain Dis **2019**, *34*, 1131-1143.

[97] Wang, X.; Ma, S.; Yang, B.; Huang, T.; Meng, N.; Xu, L.; Xing, Q.; Zhang, Y.; Zhang, K.; Li, Q.; Zhang, T.; Wu, J.; Yang, G. L.; Guan, F.; Wang, J. Resveratrol promotes hUC-MSCs engraftment and neural repair in a mouse model of Alzheimer's disease. Behav Brain Res **2018**, *339*, 297-304.

[98] Ross, C. A.; Akimov, S. S. Human-induced pluripotent stem cells: potential for neurodegenerative diseases. Hum Mol Genet **2014**, *23*, R17-R26.

[99] Lin, Y. T.; Seo, J.; Gao, F.; Feldman, H. M.; Wen, H. L.; Penney, J.; Cam, H. P.; Gjoneska, E.; Raja, W. K.; Cheng, J.; Rueda, R.; Kritskiy, O.; Abdurrob, F.; Peng, Z.; Milo, B.; Yu, C. J.; Elmsaouri, S.; Dey, D.; Ko, T.; Yankner, B. A.; Tsai, L. H. APOE4 Causes Widespread Molecular and Cellular Alterations

Associated with Alzheimer's Disease Phenotypes in Human iPSC-Derived Brain Cell Types. Neuron **2018**, *98*, 1294.

[100] Oksanen, M.; Petersen, A. J.; Naumenko, N.; Puttonen, K.; Lehtonen, S.; Gubert Olive, M.; Shakirzyanova, A.; Leskela, S.; Sarajarvi, T.; Viitanen, M.; Rinne, J. O.; Hiltunen, M.; Haapasalo, A.; Giniatullin, R.; Tavi, P.; Zhang, S. C.; Kanninen, K. M.; Hamalainen, R. H.; Koistinaho, J. PSEN1 Mutant iPSC-Derived Model Reveals Severe Astrocyte Pathology in Alzheimer's Disease. Stem Cell Reports **2017**, *9*, 1885-1897.

[101] Valadez-Barba, V.; Cota-Coronado, A.; Hernandez-Perez, O. R.; Lugo-Fabres, P. H.; Padilla-Camberos, E.; Diaz, N. F.; Diaz-Martinez, N. E. iPSC for modeling neurodegenerative disorders. Regen Ther **2020**, *15*, 332-339.

[102] Wu, Y. Y.; Chiu, F. L.; Yeh, C. S.; Kuo, H. C. Opportunities and challenges for the use of induced pluripotent stem cells in modelling neurodegenerative disease. Open Biol **2019**, *9*, 180177.

[103] Machairaki, V. Human Pluripotent Stem Cells as In Vitro Models of Neurodegenerative Diseases. Adv Exp Med Biol **2020**, *1195*, 93-94.

[104] Ochalek, A.; Mihalik, B.; Avci, H. X.; Chandrasekaran, A.; Teglasi, A.; Bock, I.; Giudice, M. L.; Tancos, Z.; Molnar, K.; Laszlo, L.; Nielsen, J. E.; Holst, B.; Freude, K.; Hyttel, P.; Kobolak, J.; Dinnyes, A. Neurons derived from sporadic Alzheimer's disease iPSCs reveal elevated TAU hyperphosphorylation, increased amyloid levels, and GSK3B activation. Alzheimers Res Ther **2017**, *9*, 90.

[105] Tcw, J. Human iPSC application in Alzheimer's disease and Tau-related neurodegenerative diseases. Neurosci Lett **2019**, *699*, 31-40.

[106] Rowland, H. A.; Hooper, N. M.; Kellett, K. A. B. Modelling Sporadic Alzheimer's Disease Using Induced Pluripotent Stem Cells. Neurochem Res **2018**, *43*, 2179-2198.

[107] Sullivan, S. E.; Young-Pearse, T. L. Induced pluripotent stem cells as a discovery tool for Alzheimers disease. Brain Res **2017**, *1656*, 98-106.

[108] Adamiak, M.; Cheng, G.; Bobis-Wozowicz, S.; Zhao, L.; Kedracka-Krok, S.; Samanta, A.; Karnas, E.; Xuan, Y. T.; Skupien-Rabian, B.; Chen, X.; Jankowska, U.; Girgis, M.; Sekula, M.; Davani, A.; Lasota, S.; Vincent, R. J.; Sarna, M.; Newell, K. L.; Wang, O. L.; Dudley, N.; Madeja, Z.; Dawn, B.; Zuba-Surma, E. K. Induced Pluripotent Stem Cell (iPSC)-Derived Extracellular Vesicles Are Safer and More Effective for Cardiac Repair Than iPSCs. Circ Res **2018**, *122*, 296-309.

[109] Liu, X. Y.; Yang, L. P.; Zhao, L. Stem cell therapy for Alzheimer's disease. World J Stem Cells **2020**, *12*, 787-802.

[110] Kang, J. M.; Yeon, B. K.; Cho, S. J.; Suh, Y. H. Stem Cell Therapy for Alzheimer's Disease: A Review of Recent Clinical Trials. J Alzheimers Dis **2016**, *54*, 879-889.

[111] Yang, H.; Xie, Z.; Wei, L.; Yang, H.; Yang, S.; Zhu, Z.; Wang, P.; Zhao, C.; Bi, J. Human umbilical cord mesenchymal stem cell-derived neuron-like cells rescue memory deficits and reduce amyloid-beta deposition in an AbetaPP/PS1 transgenic mouse model. Stem Cell Res Ther **2013**, *4*, 76.

[112] Reddy, A. P.; Ravichandran, J.; Carkaci-Salli, N. Neural regeneration therapies for Alzheimer's and Parkinson's disease-related disorders. Biochim Biophys Acta Mol Basis Dis **2020**, *1866*, 165506.

Chapter 10

The Impact of Diabetes on Hippocampus

Saeed Vafaei-Nezhad, Masood Vafaei-Nezhad, Mehri Shadi and Samira Ezi

Abstract

Maternal Diabetes is one of the most common metabolic disorders resulting an increased risk of abnormalities in the developing fetus and offspring. It is estimated that the prevalence of diabetes during pregnancy among women in developing countries is approximately 4.5 percent and this range varies between 1 to 14 percent in different societies. According to earlier studies, diabetes during pregnancy is associated with an increased risk of maternal and child mortality and morbidity as well as major congenital anomalies including central nervous system (CNS) in their offspring. Multiple lines of evidence have suggested that infants of diabetic women are at risk of having neurodevelopmental sequelae. Previous studies reveal that the offspring of diabetic mothers exhibit disturbances in behavioral and intellectual functioning. In the examination of cognitive functioning, a poorer performance was observed in the children born to diabetic mothers when compared with the children of non-diabetic mothers. Therefore, it is important to study the possible effects of maternal diabetes on the hippocampus of these infants.

Keywords: Maternal diabetes, Central nervous system, Hippocampus, Hyperglycemia

1. Introduction

1.1 Hippocampus

The hippocampus in humans is a part of the cortical region that is connected to the limbic system and consists of two cortical structures: the hippocampal formation and the parahippocampal region. Hippocampal formation refers to a group of structures with a unique cellular structure and arrangement that accompany the hippocampus and include: dentate gyrus, hippocampus, subiculum, presubiculum, and parasubiculum [1, 2]. The main difference between these two structures is the number of cortical layers and their general connections. The hippocampus in the coronal sections is a C-shaped structure located into the lower horn of the lateral ventricle. Its general shape is similar to a seahorse (**Figure 1A**) [3].

The most common classification for the hippocampus in non-human primates and other laboratory animals is, the hippocampus is divided into 4 subfields, CA1-CA4. In humans, most parts of the hippocampal formation are located on the floor of the temporal horn of the lateral ventricle (**Figure 1B**). The part of the

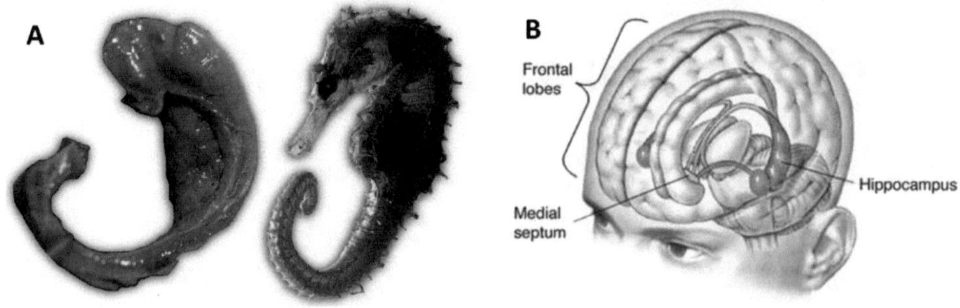

Figure 1.
Isolated hippocampus of human which is similar to seahorse (A). The shape and location of the hippocampus in humans (B).

hippocampus that is located in the floor of the lateral ventricular temporal horn (most parts of CA1 and CA2 and the distal part of CA3) is about 4 cm long [1].

1.1.1 Evolution of hippocampal formation in humans

By the ninth week of pregnancy, the primary hippocampus develops within the cerebral hemispheres but does not resemble an adult hippocampus. At the middle of the third trimester of pregnancy (weeks 19–15), immature dentate gyrus, the subiculum, and different areas of the hippocampus can be identified. The hippocampal groove deepens and different areas of the hippocampus appear to be more developed, at the end of the 25th week of pregnancy. Although cell layers are more pronounced in CA1 and CA2-CA3; But the boundary between CA1 and subiculum is not clear. By the last trimester of pregnancy (34 weeks), the hippocampal groove is narrower and the boundary between CA1 and subiculum is distinguishable; CA1, CA2, and CA3 are recognizable and seem the dentate gyrus has a mature appearance.

It should be noted that a decrease in hippocampal cell density is observed in the postnatal period, which is probably due to the apoptosis and the growth of neurons and filaments. No significant morphological changes are seen until puberty and, and the only myelination is gradually completed [4, 5].

1.1.2 Hippocampal functions

The first theory suggested that the hippocampus has a key role in olfactory functions. But this theory was not accepted because Studies in later years showed that the hippocampus did not receive any nerve fibers directly from the olfactory bulb; However, the further study indicated that the hippocampus may be involved in olfactory responses, and in particular in olfactory memory [6]. In addition, Jeffrey Gray suggested that the hippocampus might play a role in anxiety [7].

Years later, three main ideas for hippocampal function were explained: response inhibition, learning and memory, and spatial cognition [8]. The majority of psychologists and neuroanatomists believe that the hippocampus plays a principal role in the formation of new memories about experienced events (episodic or autobiographical memory), which is part of the role of the hippocampus in its activity in discovering new events, places, and stimuli [9, 10]. Some researchers believe that the hippocampus is responsible for declarative memory in addition to episodic memory [11, 12]. Severe damage to the hippocampus can cause problems with the formation of new memory, as well as impairment of earlier formed memory.

However, the memory from years before the hippocampal injury may remain intact, which appears to be due to the transfer of memory from the hippocampus to other parts of the brain over the years [8].

Interestingly, damage to the hippocampus does not affect some types of memory, such as motor memory and the ability to learn new motor and cognitive skills, such as playing a musical instrument and solving a variety of tables. This implies that these abilities depend on other types of memory called working memory, which involve different areas of the brain [13]. Several researchers distinguish between conscious recollection and familiarity which depends on the hippocampus and portions of the medial temporal lobe, respectively [14]. The hippocampus and related areas are necessary for the systematic formation and organization of memory, their retrieval, and the repetition of learned experiences. Hippocampal neurons encode a large amount of information received in the form of senses and experiences and are implicitly organized [9].

The hippocampal/internal temporal lobe (HC/MTL) complex seems to be necessary for the formation of spatial memory. This memory requires the interpretation and processing of sensory information received from the environment. In mammals in general, the proper functioning of the hippocampus, especially CA1, is essential for the formation and processing of space-related memory. Evidence suggests that the right hippocampus in humans plays a key role in spatial memory, and in rodents, the amount and accuracy of spatial memory are directly related to the number of hippocampal mossy fibers [15–17].

1.2 Diabetes

According to the World Health Organization (WHO), the term diabetes mellitus refers to a metabolic disorder with a variety of causes, including chronic hypoglycemia and impaired metabolism of carbohydrates, fats, and proteins due to impaired insulin secretion, insulin function, or both. Diabetes mellitus can have long-term effects and involve a variety of organs, including the central and peripheral nervous system, cardiovascular system, kidneys, and muscles. According to the World Health Organization, approximately 347 million people worldwide suffer from diabetes. However, 80% of these patients live in developing countries, and this number is increasing day by day [18, 19]. In addition, the number of people suffering from this metabolic disease in 2000 was 171 million, which will increase to 366 million in 2030 if proper prevention and treatment strategies are not implemented [20]. It is also estimated that by 2050 the incidence of diabetes in the world will increase by 198%, which will have a significant impact on increasing health care costs [20, 21]. As well as, global estimates suggest that by 2030, most people with diabetes will be 45 to 64 years old. The prevalence of type 2 diabetes is much faster than type 1, due to the increasing prevalence of obesity and reduced physical activity, which is one of the consequences of the industrialization of countries [22].

1.2.1 Diabetes can be divided into three general categories

1.2.1.1 Type 1 diabetes (T1D)

Type 1 diabetes, or insulin-dependent diabetes, is caused by the destruction of pancreatic beta cells as a result of insufficient insulin release. This type of diabetes is most common in adolescence and young adulthood and accounts for 10% of all diabetes cases [23].

1.2.1.2 Type 2 diabetes (T2D)

Type 2 or non-insulin-dependent diabetes, which is more common than type 1 diabetes, occurs due to insensitivity and resistance to insulin in the body along with insufficient insulin release. This type of diabetes is more common in the elderly, especially women [22].

1.2.1.3 Gestational diabetes mellitus (GDM)

Gestational diabetes mellitus is another type of diabetes that can be diagnosed during pregnancy and is defined as any amount of glucose intolerance that develops or is first diagnosed during pregnancy. However, in most cases, it is type 2 diabetes, which obviously leads to type 2 diabetes in 30 to 50 percent and in some cases has a similar course to type 1 diabetes [24]. According to this issue, diabetics and pregnant people can be divided into two groups: a group of people with diabetes who had diabetes before pregnancy (pre-existing diabetes) and may have one type of diabetes (T1D or T2D); The second group of people in whom gestational diabetes is diagnosed for the first time during pregnancy [25, 26].

1.2.2 Diabetes during pregnancy

Diabetes mellitus is the most common and important metabolic complication in pregnancy that can affect maternal and fetal health [27]. According to studies, diabetes is seen in around 7% of pregnancies and its prevalence depends on the study population and diagnostic tests from 1 Up to 14% have also been reported [24, 28]. Gestational diabetes is one of the leading causes of mortality in pregnant women which can be elevating the risk for spontaneous abortion, stillbirth, congenital malformations, and perinatal morbidity and mortality [29]. It is well documented that maternal glycemic control during pregnancy can markedly decrease congenital malformation outcomes in the fetus. Studies have shown that infants born to diabetic mothers have a higher risk of congenital disorders in the nervous, cardiovascular, kidney, and gastrointestinal tracts [26, 30–32].

1.2.2.1 Pathophysiology of gestational diabetes on embryonic development

In healthy mothers and under normal conditions, pregnancy causes hyperplasia of pancreatic beta cells and increases insulin levels in the mother's bloodstream [33]. On the other hand, at the beginning of pregnancy, insulin sensitivity is observed in pregnant women, which turns into insulin resistance as the pregnancy progresses. Maternal insulin resistance appears to occur due to the production of placental diabetogenic hormones such as growth hormone, placental lactogen, corticotropin-releasing hormone, and progesterone [33, 34]. This insulin resistance decreases after the placenta leaves the mother's body and increases the risk of hyperglycemia in mothers 7 to 15 weeks after delivery [35, 36].

Previous studies have illustrated that increase in the level of maternal blood glucose and a decrease in insulin is the main reason for diabetes during pregnancy [37]. In the above conditions, glucose can easily pass through the placenta into the fetal bloodstream, leading to fetal hyperglycemia. During the first few weeks of pregnancy, fetal islet cells (beta cells) cannot release enough insulin in response to hyperglycemia, which is the main cause of fetal hyperglycemia. In response to this condition, after week 20, the fetal pancreas is stimulated and the pancreatic beta cells begin to hypertrophy and hyperplasia, which eventually leads to increased fetal insulin levels. In addition to impairing the development of various organs, this complication can be followed by hypoglycemia and hyperinsulinemia in the first few days after birth [30, 37–39].

Results from previous experiments have shown that insulin can influence carbohydrate, fats, and protein metabolism, membrane transport of glucose, amino acids, and ion exchange in cells as well as protein and DNA synthesis. In addition, insulin can stimulate or inhibit the activity of certain enzymes and regulate gene expression [40]. On the other hand, alters and reduction of ions transfer can lead to the reduction level of some vital ions such as zinc. Thereby, this process has a negative effect on the migration of marginal layer cells in the fetus of diabetic mothers which increases defects in the central nervous system [41]. In this regard, some studies indicate that high concentrations of beta-hydroxybutyric acid, which occurs in diabetes mothers, can delay the development of the central nervous system of the fetus [42–44].

Although hyperglycemia is believed to be the most important teratogenic element in diabetic pregnancy; Some researchers suggest that changes in maternal metabolic status (i.e., triglyceride and β-hydroxybutyrate levels and branched-chain amino acids) lead to disrupted fetal metabolism of inositol, sorbitol, prostaglandins, and arachidonic acid could have a teratological effect and therefore be important for the incidence of fetal disorders. An excess of fetal reactive oxygen species (ROS) has also been linked to the etiology of congenital malformations induced by diabetes. These free radicals may cause increasing neuronal death by oxidizing proteins, damaging DNA, and inducing the lipoperoxidation of cellular membranes. In vitro and in vivo studies have shown that the disturbed development of embryos in a diabetic milieu can be normalized by treatment with different antioxidant factors [27, 45–47].

1.2.3 The effects of gestational diabetes on fetal development and infant health

It is well documented that fetuses of mothers with diabetes during pregnancy are in a completely different environment than a healthy mother. Glucose, alanine, and free fatty acids are transported in large quantities from the mother's blood to the fetus. As a result, the concentration of insulin in the amniotic fluid increased, which indicates a compensatory response of the fetus to an increase in these factors [48]. Hyperglycemia in the first trimester of pregnancy increases significantly the risk of congenital malformations and stillbirth [49].

Several studies have shown that maternal hyperglycemia during pregnancy causes fetal hyperglycemia and neonatal hypoglycemia; Because circulating glucose simply crosses the placenta by facilitating diffusion, resulting in fetal hyperglycemia [35, 37]. In contrast, to compensate for this event, the fetal pancreas is stimulated and the pancreatic beta cells begin to hypertrophy, which causes increased fetal insulin levels. Due to the inability of the growing beta cells in the pancreas to secrete enough insulin, this condition soon leads to fetal hypoinsulinemia [37, 50]. Although this complication is temporary; However, studies show that fetuses from mothers with diabetes develop hyperinsulinemia in the last trimester of pregnancy. This condition in the fetus, in addition to affecting various organs, puts infants at risk for hypoglycemia in the few first days after birth, which is one of the most important causes of infant mortality in diabetic mothers [51, 52]. Previous studies report that gestational diabetes can increase the risk of impaired fetal and neonatal development, mortality, and also problems in infancy, childhood, and adulthood [31, 32]. Abundant human studies have identified type 1 diabetes during pregnancy as an important factor in the development of fetal and neonatal complications such as stillbirth, fetal macrosomia, respiratory distress syndrome, diabetes, jaundice, asphyxia, hypertension, neonatal hyperglycemia, hypocalcemia and hypomagnesemia, cardiac abnormalities, hypoxia, and neonatal polycythemia [53–55].

Studies have also shown that diabetes can have teratogenic effects and also negative effects on embryogenesis, organogenesis, and fetal growth [56]. The frequency of the mentioned problems is the same for both types of diabetes and the incidence of these complications depends directly on the severity of maternal diabetes [57]. Studies have shown that in gestational diabetes, there is a linear relationship between maternal glucose levels in early pregnancy and the incidence of birth defects [58].

1.2.3.1 Fetal and infant mortality

In past years, the rate of infant mortality from diabetic mothers has been significantly high. But nowadays, due to advances in medical and obstetrical management in the pregnancy period, this rate has decreased significantly. But despite these advances, the mortality rate in these infants is still reported to be 3 to 10 times higher than in infants under normal pregnancy conditions, as well as the prevalence of congenital malformations is 4 to 10 times higher than that of healthy mothers [29, 33]. Studies show that fetal malformations are the reason 30 to 40 percent of infant deaths in diabetic mothers. However, it has been clearly shown that precise control of blood glucose during pregnancy reduces maternal and neonatal mortality [59, 60].

1.2.3.2 Neonatal hypoglycemia

A sharp drop in plasma glucose concentration after delivery is a characteristic feature of newborns born to poorly controlled diabetic mothers. This event occurs due to chronic maternal hyperglycemia resulting in fetal pancreatic cell hyperplasia. Subsequently, this hyperplasia causes stimulation of fetal pancreatic beta cells to release a high level of insulin. In addition to stimulating somatic growth, hyperinsulinemia is also one of the main causes of hypoglycemia in the first few minutes after birth [24, 49].

1.2.3.3 Congenital malformations

Fetal malformations, which usually occur between 7 and 10 weeks, account for 30 to 40 percent of prenatal deaths. Studies show that the rate of severe malformations in children of healthy mothers is 7.8%, but 15% in children of diabetic mothers. These malformations usually affect the central nervous system, heart, kidneys, and urinary system [27, 33].

1.2.3.4 Macrosomia

It is more common in infants of mothers with diabetes and increases the risk of death at birth. Diabetes during pregnancy can double the incidence of macrosomia and other neonatal anthropometric indexes in diabetic mothers compared to babies born to non-diabetic mothers. Previous studies have clearly established that maternal diabetes can induce macrosomia in most fetal organs except the brain. These events are closely related to fetal hyperinsulinemia and maternal hyperglycemia [24, 61].

1.2.3.5 Neuropsychological effects of gestational diabetes on infants

Previous studies have shown that infants of diabetic mothers have a significant decrease in brain weight and size compared to healthy mothers [37]. Moreover, multiple lines of evidence indicated that offspring of diabetic mothers exhibit disturbances in behavioral and intellectual functioning. In this way, other studies have shown a link between brain size and intelligence. These children also show poorer cognitive function than children of healthy mothers, which is related to the effects of maternal metabolic changes on the development of the fetal central nervous system (CNS) [62–64]. Long-term studies of offspring born to diabetic mothers have shown that diabetes during pregnancy reduces IQ in these children [65].

Researchers believe that gestational diabetes is a teratogen for the development of the central nervous system [66]. Since the brain is one of the major organs using up glucose in the body, any defect in the process of supplying glucose to the brain, even for a short time, can cause brain disorders [67]. In humans, the differentiation and maturation of the cerebral cortex occur at the end of the second trimester of pregnancy, and therefore at this time, any change in blood glucose levels can have irreversible effects [65, 67, 68]. It is believed that hypoglycemia due to hyperinsulinemia in infants born to mothers causes neuronal damage in the internal temporal region and also memory-related areas [69].

Smoak and Sadler examined the role of glucose in brain development in mouse embryos. The researchers showed that a 50 percent reduction in blood glucose levels for 2 hours in mice at the stage of neural tube formation causes developmental disorders in the nervous system [70]. Habituation is a sign of proper functioning of the central nervous system. Studies have shown that fetuses of diabetic mothers have poorer habituation than fetuses of healthy mothers. This reduction is indicative of the effect of gestational diabetes on suitable central nervous system function [62, 71].

1.2.3.6 Attention deficit disorder and hyperactivity

Neural damage in infants of mothers with diabetes during pregnancy is not only limited to a negative effect on their intelligence but also can reduce their concentration [65]. Studies have indicated a higher incidence of developmental delay and behavioral problems including short attention span, over-activity, and attention-seeking in children born to mothers with diabetes [58]. Moreover, growth motor skills and speech and language delay were the main development areas of concern that could link between maternal diabetes and development in children aged 1–6 years [72]. Other studies demonstrated that school-age children younger than 9 years, born to diabetic mothers, had a higher rate of attention deficit, lower cognitive scores, and lower gross and fine motor achievements than matched control children did. as well as the period might affect the later cognitive and behavioral function of progeny by influencing developing brain cells in utero [73, 74].

In general, it can be said that diabetes during pregnancy disrupts the development, function, and maturity of the CNS in the children of diabetic mothers, which manifests itself in the form of intelligence, educational and behavioral problems. It should be noted that both increase and decrease in blood sugar during brain development can lead to a decrease in various cognitive functions [64, 65, 75].

1.2.3.7 Intelligence and memory disorders

Children born to diabetic mothers suffer from neuropsychiatric disorders, including decreased intelligence and memory disorders. In 1997, Rizzo and colleagues followed 139 women with gestational diabetes and reported that their children's intelligence was directly related to maternal glucose metabolism in the second and third trimesters of pregnancy [76]. In animal studies, the hippocampus, which is particularly involved in memory, has been identified as the most susceptible area to the negative effects of hypoglycemia, which can occur in the fetus or immediately after birth. Overall, the results of various studies indicate the negative impact of maternal diabetes on cognitive functions, which can cause memory and intelligence deficits [75, 77, 78].

1.3 Effects diabetes during pregnancy on hippocampus

As mentioned before, the hippocampus is an important brain structure crucial for spatial learning and memory. In diseases that cause memory loss and other intellectual functions, such as Alzheimer's disease, hippocampus cells are among the first cells to undergo degenerative changes. it is well documented that the hippocampus provides a stimulus that converts short-term memory to long-term memory, and whatever its mechanism, it would not have happened without the hippocampus [3].

Studies have shown hippocampus is very sensitive to changes in glucose concentration during development. in this regard, there is a bulk of studies that show children born to diabetic mothers are more likely to have neurodevelopmental abnormalities including impairments in memory, learning ability, activity level, attention span, and motor functioning. These infants also show lower IQ scores compared to infants born to healthy mothers [37, 58].

When pregestational and gestational diabetes-exposed children were grouped together in the study of DeBoer et al., it was demonstrated a negative link between maternal diabetes and development of memory, circuitry, and behavioral mnemonic performance in children at 1-year of age. Moreover, they showed that the metabolic abnormalities due to diabetes during pregnancy alters prenatal development, which can influence memory performance on a delay recall task. It is well documented that metabolic abnormalities which occur in diabetes during pregnancy can impair fetal CNS development, which leads to structural and functional defects, especially in the hippocampus [79].

Experimental models of diabetes during pregnancy in animals have shown a decrease in the numerical density of neurons in some parts of the fetal CNS, particularly in the hippocampus, which is reflected in decreased memory and learning and impaired memory storage and recall of information [37].

In the study by Sadeghi et al., stereological change in the hippocampus of rat offspring due to diabetes in pregnancy was evaluated. In that study, the authors found a significant reduction in total volumes of the hippocampus in offspring born to diabetic mothers when compared to the control group. In addition, their results have been shown the hippocampal subfields volumes, especially the CA1, DG, and subiculum, were significantly decreased. Moreover, they reported a significant decrease in the number of hippocampal cells in infants born to diabetic mothers [80].

A study by Tehranipour and Khakzad examines the effect of maternal diabetes on neural density in the hippocampus of newborn rats immediately after birth. Their results showed that diabetes during pregnancy can decrease the number of neurons in the hippocampus, especially in the CA3 area [81].

Synaptogenesis is one of the key events which happen throughout the development of the central nervous system. The chemical synapses in the central nervous system contain the presynaptic apparatus, the synaptic cleft, and the postsynaptic region. During synaptic transmission, In the presynaptic part, neurotransmitters stored in the synaptic vesicle through the exocytosis process, release into the synaptic cleft and then fuse with their receptors in the postsynaptic membrane. This process is highly regulated in CNS. The bulk of studies were performed to recognize and specify the components of the synaptic vesicle membrane. These studies have been found several proteins, including synaptophysin, synaptobrevin, and synaptogamin that functions as the regulators of exocytosis. In recent studies, synaptophysin has been utilized as a valid marker for synaptic density and synaptogenesis [66, 82, 83].

In a study by Vafaei-Nezhad et al., researchers revealed that the SYP expression levels were significantly reduced in hippocampus sub-regions of pups born to diabetic animals, especially at P7 and P14, compared with the control group [66]. SYP as a major protein of the synaptic vesicle membrane may play an important role in transmitter release. Thus, the early decrease in SYP expression may reflect a down-regulation of synaptic functions and may be related to the release of the neurotransmitters. Since SYP and other synaptic vesicle proteins have been implicated in the mechanisms of cellular plasticity underlying learning, a decrease in the expression of this protein might disrupt memory formation [84–86]. SYP is also a reliable indicator of synaptic plasticity, and has previously been demonstrated to correlate well with the loss of cognitive function in animal models with neurodegeneration and in humans with Alzheimer's disease [87]. Earlier study also demonstrated a correlation between aging-related deficits in cognitive functions and disturbances in SYP expression in the hippocampus. There are also documents showing that upregulation of SYP expression may contribute to the mechanisms that underlie learning and memory [85, 86].

In another investigation by Sadeghi et al., the effects of maternal diabetes on neurogenesis in the developing hippocampus were examined. in the study, researchers probe NeuN and DCX markers changes. They found a significant higher mean number of DCX-positive cells and an up-regulation in mRNA expression of DCX in neonates born to diabetic mothers. Moreover, they demonstrated a significant reduction in the mean number of NeuN-positive cells and down-regulation in NeuN expression in the newborns to diabetic animals [88].

This author in another study also revealed that maternal diabetes could result in developmentally induced increase in the hippocampal GFAP expression and numerical density of GFAP positive cells in the DG hippocampal subfield of the offspring born to diabetic mothers compared to healthy mothers [89].

In experimental animals, it is reported that diabetes during pregnancy can increase apoptosis in the central nervous system in neonates. In this context, research by Lotfi et al. revealed that maternal diabetes in newborns of diabetic mothers leads to a marked increase in the number of apoptotic cells in the CA3 subregion of the hippocampus. They also suggested that hyperglycemia during pregnancy could cause a developmental change in the density of neuronal cells in the offspring's hippocampus [90]. In consist with this study, in another research has been reported that diabetes during pregnancy leads to up-regulate in Bax and down-regulated in Bcl-2 gene expression (apoptosis-regulatory genes) in the hippocampus of rat neonates born to mothers with diabetes [91].

Nowadays, neurotrophic factor such as brain-derived neurotrophic factor (BDNF) is considered for its role in regulating the development of the fetal nervous system. In the experiment by Sardar et al., they assessed the effects of maternal diabetes on gene expression and distribution pattern of BDNF in the hippocampus of neonatal rats. The results of the present study showed that diabetes during pregnancy causes marked BDNF downregulation and numerical density of $BDNF^+$ cells reduction in both sides' hippocampi of the male/female diabetic group [92].

In another study by Hami et al., the effects of diabetes in pregnancy on gene expression and protein concentration of IGF1R and IR in the developing rat hippocampus at postnatal days 0, 7, and 14 were evaluated. In that study, the authors found a markedly upregulation of both IR and IGF1R expression in the hippocampus of diabetic group newborns at first postnatal day. At the same time point, they showed only slight changes in their hippocampal protein transcripts. In 7-day, old rats, there was a significant decreased in IGF-1R gene expression and protein levels in the newborns born to diabetic dams. Moreover, they found a down regulation in hippocampal IGF1R transcripts in 14-day old diabetic group offspring. Two weeks after birth, the IR gene expression was significantly declined in the hippocampus of diabetic newborns [18].

1.4 Effects diabetes on the hippocampus in adults

A bulk of studies have been demonstrated that all types of diabetes have adverse effects on the central nervous system such as disruption in hypothalamic and hippocampal neuropeptides gene expression, a change in hippocampal function and decreased hippocampal synaptic plasticity, glutamate neurotransmission abnormalities, and neurotoxicity [93, 94]. Studies have shown that people with diabetes have a higher risk of developing memory and learning disorders and also Alzheimer's disease, depression, stroke, and dementia compared to healthy people [95]. A previous study showed that diabetes induction in animal models can elevate corticosterone levels and defects hippocampal synaptic plasticity, learning, and long-term potentiation in the CA1-field [96, 97]. Recent studies showed that STZ-induced diabetes notably decreased the number of proliferating cells in the dentate gyrus of rats by changing the hippocampal synaptic plasticity [98].

There is some evidence that diabetes can decrease hippocampal cell proliferation and survival, while some researchers suggest that diabetes also has adverse effects on neuronal differentiation. Several preclinical studies revealed the bulk of evidence that diabetes has adverse effects morphological integrity of the hippocampus and that it reduced hippocampal neurogenesis. These hippocampal dysfunctions may cause cognitive and mood disorders in diabetic's people [99].

In the study by Revsin et al., Neuronal and astroglial alterations in the hippocampus of a mouse model for type 1 diabetes have been evaluated. Their results showed that diabetic condition in mice can cause a significant increase in the number of astrocytes positive for apolipoprotein-E (Apo-E), a marker of ongoing neuronal dysfunction, abnormal expression of Jun + neurons in CA1 and CA3 layers and dentate gyrus, and Fos-expressing neurons in CA3 layer, and augmented activity of NADPH-diaphorase which is linked to oxidative stress, in CA3 region. They state that these changes could be one of the reasons for the negative effects of diabetes on the hippocampus [94]. Similar to the results of this study Stereological studies have shown that there is no significant change in the total neuron number values and the volume of the hippocampus in diabetic animals compared to healthy animals [94, 100].

It is well known that type 2 diabetes mellitus can lead to cognitive deficits in patients. In this regard, a study by Liu et al., has been revealed that diabetes could alter effective connectivity (EC) between hippocampus and default mode network (DMN), which is interpreted to be related to cognitive disorders in patients with T2DM especially affecting learning and memory [101].

In experimental animals by Kamal et al., they assessed Synaptic transmission changes in diabetic rat s' hippocampus. Their data illustrated that intracellular recording from the pyramidal hippocampal cells of the AMPA summation responses in the diabetic animals was markedly lower than control animals. Hence, they suggest that diabetes can change pre and postsynaptic cells functions which may play a key role in in the synaptic plasticity disorder observed in diabetic animals [93].

Defects in glucose metabolism in the nervous system are not only important for neural cells function but also normal astrocyte activity is related to it. Astrocytes have many important roles in central nervous system such as glucose uptake. Astrocyte dysfunction in the hippocampus have been seen in pathological conditions such as aging, stress, autoimmune diseases, neurodegenerative diseases, and diabetes. In the hippocampus of the animal diabetic model has been reported a significant increase in GFAP immunoreactivity, glial fibrillary acidic protein (GFAP), an astrocyte intermediate filament cytoskeletal protein which is considered the main indicator of astroglial activation caused by CNS injury [102]. In line with this study, a study by Lebed et al. showed that diabetes can alter GFAP and S100B levels in the hippocampus. These findings propose that the reaction of astroglial cells can be the first reaction to impaired glucose metabolism, which is likely to play a crucial role in the mechanisms underlying diabetes-related disorders of CNS function [103].

In a study by Pamidi et al., the Effect of streptozotocin-induced diabetes on rat hippocampus was assessed. They used Cresyl violet staining for evaluating the number of surviving neurons in the subfield of the hippocampus. Their findings revealed that the number of survived neurons in the subfield of the hippocampus (CA1, CA2, CA3, dentate hilus, dentate gyrus) in diabetic animals was significantly reduced compared to controls. Hence, researchers suggested that uncontrolled and long-term diabetes could cause intense hippocampal neurodegeneration [104].

In another study, researchers examined the effects of STZ-induced diabetes on NCAM protein expression in various parts of the nervous system, including the hippocampus. The results of this study have been demonstrated that diabetic animals developed remarkable defects in learning and memory behaviors which were evaluated by passive avoidance and water maze tests. Their results also indicate that streptozotocin-induced diabetes disrupts cognitive functions and causes an imbalance in NCAM expression in some parts of the brain, particularly the hippocampus, which is involved in memory and learning. Based on the results of their study, the researchers concluded that changes in the expression of NCAM in the hippocampus of diabetic people could be one of the main causes of memory and learning disorders [105].

Axonal transport has a critical role in normal CNS function and disruption of this process has been linked to various neurodegenerative diseases and could also play a role in diabetes-related diseases that affect the nervous system. Given the importance of this process, Baptista et al. assessed the impact of diabetes on axonal transport in the hippocampus. in this regard, they evaluated KIF1A, KIF5B, and dynein in the hippocampus. Their results showed a significant increase in KIF1A and KIF5B mRNA and protein levels, in CA1, CA3, and DG hippocampal subregion, in the diabetic animals. Nevertheless, no changes in dynein protein were observed [106].

The animal experiment has been revealed that diabetes can cause a reduction of nuclear and perikaryon diameters as well as neuronal density in the sub-region of the hippocampus (CA1, CA2, CA3, and dentate gyrus) [107]. In addition, it has been reported that levels of Caspase-3, Bax, and Bcl-2 in the hippocampus increase after diabetes [108].

1.5 Effects of diabetes during pregnancy on the other parts of CNS

The source of the mammalian neural tube is the ectoderm layer in the embryo. During the first few months of pregnancy which is called the embryonic period, part of the ectoderm becomes a neural plate. Then, this plate is folded, raised, and connected in the midline, which becomes the neural tube [109, 110]. The mentioned procedure is mediated by several signaling molecules and transcription elements [111, 112]. There is strong evidence that diabetes during pregnancy causes abnormalities in the development of the nervous system, including the neural tube, which itself can cause neural tube defects (NTD) [113, 114]. Neural tube formation during the embryogenesis period is tightly regulated by several cellular mechanisms such as proliferation, migration, differentiation, and apoptosis of neural progenitor cells. Previous studies have shown that pregnancy-related hyperglycemia can disrupt cell migration and proliferation, as well as induce programmed cell death during embryogenesis. Studies have revealed that any changes in the mentioned processes can cause NTDs [95, 115].

Spina bifida, which is one of the common NTD, results from failure of neural tube fusion in the spinal area. This abnormality is a common congenital defect that occurs in infants of diabetic mothers [116]. The exact mechanism underlying the teratogenic effects of diabetes during pregnancy is not entirely understood, but it is suggested that the incidence and intensity of congenital malformations in infants born to diabetic mothers are correlated with the degree of maternal metabolic control [37].

Diabetes during pregnancy causes biological changes in the mother that can affect the development of the fetal nervous system. These biological changes can affect nervous system development and neurological abnormalities in the fetus by affecting neurotransmitters, synaptic membranes, and the expression of growth factors that are involved in nervous system development. Studies have shown that increased oxidative stress plays an important role in fetal development and indirectly can cause some of the nervous system developmental disorders that occur in fetuses from diabetic mothers [22]. Free radicals can also alter the biological activity of proteins and peptides, increase neuronal death, and cause DNA damage [37, 117, 118]. Experimental research showed that maternal hyperglycemia can lead to a significant decrease in the volume of gray and white matter, and also can reduce the number of neurons in the gray matter of the CNS in this offspring [119].

Evaluations of cognitive functioning and the behavior of the children born to diabetic mothers provide the chance to functionally assess the CNS development. Hence, behavioral and cognitive assessments in neonates of diabetic mothers can shed light on the effects of maternal diabetes on CNS development [64, 65, 72, 73].

Animal models of diabetes during pregnancy indicate a decrease in the numerical density of neurons in some parts of the fetal central nervous system, particularly in the brain, hippocampus, and cerebellum, due to diabetes during pregnancy, which may be followed by decreased memory and learning ability, and memory defects and information retrieval [37, 58, 80, 120]. Studies by Khaksar on gestational diabetes have shown that gestational diabetes can affect the infant CNS and reduce the number of cells and the thickness of the white and gray matter of the infant cerebrum, cerebellum and spinal cord [119]. Earlier research on the neurologic development in infants of diabetic mothers illustrated fundamental CNS deficits even when there were no structural abnormalities. These alterations were markedly less serious when maternal diabetes was controlled and treated, but some alterations in cognitive function may continue all over childhood [37].

The study by Hami et al. showed that diabetes during pregnancy in infants born to diabetic mothers can lead to a significant decrease in cerebellar volume, the thickness of cerebellar cortical layers, and also decreases in the numerical densities of cerebellar Purkinje and granular cells. They suggested that this event may delay the normal development of the cerebellum and may be a cause of the motor, behavioral, structural, and cognitive disorders seen in the offspring of diabetic mothers [120].

It is well documented that diabetes during pregnancy has structural and functional effects the on central nervous system in both human and animal studies. Despite much research in the field, the exact mechanism by which diabetes affects the development of the nervous system during pregnancy is not yet clearly understood. Some studies have suggested that diabetes itself may have teratogenic effects that can increase the risk of abnormal fetal development in the fetal organs [37].

Some studies suggested that hyperglycemia conditions during pregnancy can induce programmed cell death and impair cell proliferation in mouse embryo neural tubes. Recent experiments suggest that tumor necrosis factor (TNF) plays an important role in neurodevelopmental disorders in babies born to diabetic mothers. TNF could cross the placenta and enter the embryo's bloodstream, which has neurotoxic effects on fetal brain development and can also lead to white matter damage and cerebral palsy [35, 121].

Other investigations implicated that infants born to diabetic mothers have iron metabolism abnormalities. This deficiency can lead to neurodevelopmental and neurobehavioral disorders. Experimental studies in animal models have shown that iron impairment during pregnancy and after childbirth has negative effects on myelination, the metabolism of neurotransmitters in the brain, and the regulation of brain energy. The perinatal iron disorder can elevate the vulnerability of the neonatal brain, especially the hippocampus, to the hypoxic–ischemic insult which leads to abnormal cognitive processing in the newborn period [122, 123].

Kinney *et al*. found that the only female offspring born to diabetic dams showed deficits in long-term memory and learning. These results have suggested that the *in utero* diabetic condition has gender-specific effects on CNS development [78]. In a study by Plagemann *et al*., alterations in catecholamines levels in the hypothalamic nuclei of newborns born to diabetic animals were evaluated. They reported an increased hypothalamic dopamine (DA) and norepinephrine (NE) concentrations in the offspring born to diabetic rats at birth. Twenty-one day- old pups born to diabetic mothers, NE levels were strikingly increased in the ventromedial hypothalamic nucleus and the lateral hypothalamic area (LHA), while DA levels were significantly elevated in the paraventricular hypothalamic nucleus and the LHA. The authors concluded that there are strikingly differences in hypothalamic catecholaminergic systems during early development in the rat newborns born to diabetic animals [124].

Studies have shown that brain weight in infants born to diabetic mothers is significantly lower than in infants born to healthy mothers [66]. Interestingly, Xiang et al. explained a close correlation between diabetes during pregnancy with an elevated risk of autism spectrum disorder (ASD) in infants [125]. Animal models of diabetes during pregnancy indicate a decrease in the numerical density of neurons in some parts of the fetal central nervous system, particularly in the brain, hippocampus, and cerebellum, due to diabetes during pregnancy, which may be followed by decreased memory and learning ability, and memory defects and information retrieval [37, 58].

Much research has shown that fetal hyperglycemia during pregnancy can alter gene expression that is involved in the proliferation and differentiation of nerve cells. These changes can be the basis of the neurocognitive and neurodevelopmental disorders seen in babies of diabetic mothers [37, 58].

Previous studies have shown that insulin and insulin receptor (InsR), and insulin-like growth factor-1 (IGF-I) and IGF-1 receptor (IGF-1R) play key roles in regulating the growth and development of the CNS. It is well documented that insulin and IGF-1 stimulate the proliferation of neuronal progenitor cells, increases the survival of neurons and oligodendrocytes, elevate synaptogenesis and neurotic outgrowth, inhibit neural cell apoptosis, and induces differentiation of neurons. With this in mind, the researchers examined the effects of maternal diabetes on the expression of this factor in the nervous system. Their results have been revealed that maternal diabetes can strongly influence the regulation of these factors at the level of mRNA and protein in neonates' central nervous system [18, 37, 58, 126, 127].

Synaptic transmission and information transfer are highly regulated processes in the nervous system. Information transfer occurs when neurotransmitters stored in synaptic vesicles release into the synaptic cleft and attach to their receptors in the postsynaptic cell membrane. Synaptic vesicles are responsible for collecting neurotransmitters and releasing them into the synaptic cleft through exocytosis. The exocytosis process during the release of neurotransmitters has been extensively studied [82, 83]. Several families of proteins that are present in the membranes of synaptic vesicles which involved in the regulation of this process have been identified [82]. One of these proteins is synaptophysin. This protein is widely found in the membranes of vesicles containing neurotransmitters in neurons.

Synaptophysin is a protein involved in the construction of synaptic vesicles that researchers use as a suitable marker of synaptic density, synaptogenesis [128]. Studies have also shown that synaptophysin may be involved in the exocytosis of synaptic vesicles [129]. It is well documented that synaptophysin levels also change in pathological brain conditions such as Alzheimer's, Parkinson's, schizophrenia, and bipolar disorder [87, 130–132]. Previous studies illustrated that diabetes during pregnancy can reduce significantly downregulate synaptophysin gene expression in different layers of the neonate cerebellum which is born to diabetic mothers [74].

2. Conclusions

As the prevalence of diabetes increases in various societies, particularly in developing countries, the total number of fetuses born to diabetic mothers will continue to increase in the coming decades. The metabolic changes caused by maternal diabetes disrupt glucose homeostasis in the fetus and cause numerous problems for the fetus. Nowadays, these metabolic disorders are discussed in various studies. Previous studies have clearly shown an association between maternal blood sugar levels and an increased risk of birth defects. Researchers have found that maternal hyperglycemia can have teratogenic effects on fetuses. However, the exact mechanism of the cause of the malformations in the fetuses of diabetic mothers is not yet known. It is well documented that diabetes during pregnancy has significant effects on fetal central nervous system development, particularly the hippocampus, which can lead to disorders in learning, memory, and attention in newborns and adults. In this regard, it is suggested that suitable management of hyperglycemia and dissecting out the mechanisms responsible for diabetes-related changes in the functions of the hippocampus, could help to prevent impaired cognitive and memory functions in offspring and adults.

Conflict of interest

The authors declare no conflict of interest.

Acronyms and abbreviations

CNS	Central nervous system
WHO	World Health Organization
T1D	Type 1 diabetes
T2D	Type 2 diabetes
GDM	Gestational diabetes mellitus
ROS	Reactive oxygen species

NTD	Neural tube defects
TNF	Tumor necrosis factor
DA	Dopamine
NE	Norepinephrine
LHA	Lateral hypothalamic area
ASD	Autism spectrum disorder
InsR	Insulin receptor
IGFI	Insulin-like growth factor-1
IGF-1R	IGF-1 receptor
HC/MTL	Hippocampal/internal temporal lobe
Apo-E	Effective connectivity (EC)
DMN	default mode network

Author details

Saeed Vafaei-Nezhad[1,2]*, Masood Vafaei-Nezhad[2], Mehri Shadi[3] and Samira Ezi[4]

1 Department of Anatomical Sciences, School of Medicine, Birjand University of Medical Sciences, Birjand, Iran

2 Cardiovascular Diseases Research Center, Birjand University of Medical Sciences, Birjand, Iran

3 Tissue Engineering Lab, Department of Anatomical Sciences, School of Medicine, Shiraz University of Medical Sciences, Shiraz, Iran

4 Faculty of Medicine, Department of Anatomy, Gonabad University of Medical Sciences, Gonabad, Iran

*Address all correspondence to: vafayisaeed@gmail.com

References

[1] Roelink H. Hippocampus formation: an intriguing collaboration. Current Biology. 2000;10(7):R279-RR81.

[2] Rosene DL, Van Hoesen GW. The hippocampal formation of the primate brain. Cerebral cortex: Springer; 1987. p. 345-456.

[3] Teyler TJ, Discenna P. The topological anatomy of the hippocampus: a clue to its function. Brain Research Bulletin. 1984;12(6):711-719.

[4] Arnold SE, Trojanowski JQ. Human fetal hippocampal development: II. The neuronal cytoskeleton. Journal of Comparative Neurology. 1996;367(2):293-307.

[5] Humphrey T. The development of the human hippocampal fissure. Journal of anatomy. 1967;101(Pt 4):655.

[6] Moser MB, Moser EI. Functional differentiation in the hippocampus. Hippocampus. 1998;8(6):608-619.

[7] Gray JA, McNaughton N. The neuropsychology of anxiety: reprise. 1996.

[8] Bird CM, Burgess N. The hippocampus and memory: insights from spatial processing. Nature Reviews Neuroscience. 2008;9(3):182-194.

[9] Nadel L, Moscovitch M. Memory consolidation, retrograde amnesia and the hippocampal complex. Current opinion in neurobiology. 1997;7(2):217-227.

[10] Moscovitch M, Rosenbaum RS, Gilboa A, Addis DR, Westmacott R, Grady C, et al. Functional neuroanatomy of remote episodic, semantic and spatial memory: a unified account based on multiple trace theory. Journal of anatomy. 2005;207(1):35-66.

[11] Umbach G, Kantak P, Jacobs J, Kahana M, Pfeiffer BE, Sperling M, et al. Time cells in the human hippocampus and entorhinal cortex support episodic memory. Proceedings of the National Academy of Sciences. 2020;117(45):28463-28474.

[12] Pushp S, Kashmira P, Hazarika SM. Declarative Memory-based Structure for the Representation of Text Data. arXiv preprint arXiv:200210665. 2020.

[13] Liu J, Chen Y, Stephens R, Cornea E, Goldman B, Gilmore JH, et al. Hippocampal functional connectivity development during the first two years indexes 4-year working memory performance. Cortex. 2021;138:165-177.

[14] Squire LR. Mechanisms of memory. Science. 1986;232(4758):1612-1619.

[15] Olton DS, Becker JT, Handelmann GE. Hippocampus, space, and memory. Behavioral and Brain sciences. 1979;2(3):313-322.

[16] Howard MW, Fotedar MS, Datey AV, Hasselmo ME. The temporal context model in spatial navigation and relational learning: toward a common explanation of medial temporal lobe function across domains. Psychological review. 2005;112(1):75.

[17] Eichenbaum H, Dudchenko P, Wood E, Shapiro M, Tanila H. The hippocampus, memory, and place cells: is it spatial memory or a memory space? Neuron. 1999;23(2):209-226.

[18] Hami J, Sadr-Nabavi A, Sankian M, Balali-Mood M, Haghir H. The effects of maternal diabetes on expression of insulin-like growth factor-1 and insulin receptors in male developing rat hippocampus. Brain Structure and Function. 2013;218(1):73-84.

[19] Konishi J, Misaki T, Saga T, Miki Y. Progress in studies of endocrinology and metabolism in the field of internal medicine in the last 100 years: Diagnostic imaging of endocrine diseases. Nihon Naika Gakkai zasshi The Journal of the Japanese Society of Internal Medicine. 2002;91(4): 1117-1121.

[20] Shaw JE, Sicree RA, Zimmet PZ. Global estimates of the prevalence of diabetes for 2010 and 2030. Diabetes research and clinical practice. 2010;87(1):4-14.

[21] Wild S, Roglic G, Green A, Sicree R, King H. Global prevalence of diabetes: estimates for the year 2000 and projections for 2030. Diabetes care. 2004;27(5):1047-1053.

[22] Quinn L. Type 2 diabetes: epidemiology, pathophysiology, and diagnosis. The Nursing Clinics of North America. 2001;36(2): 175-192, v.

[23] Daneman D. Type 1 diabetes. The Lancet. 2006;367(9513):847-858.

[24] Larijani B, Hossein-nezhad A. DIABETES MELLITUS AND PREGNANCY. Journal of Diabetes and Metabolic Disorders. 2002;1(1):9-22.

[25] Jelsema RD. Management of diabetes mellitus complicating pregnancy. Obstetrics & Gynecology. 2004;103(3):586.

[26] Allen VM, Armson BA, Wilson RD, Blight C, Gagnon A, Johnson J-A, et al. Teratogenicity associated with pre-existing and gestational diabete. Journal of Obstetrics and Gynaecology Canada. 2007;29(11):927-934.

[27] Schwartz R, Teramo KA, editors. Effects of diabetic pregnancy on the fetus and newborn. Seminars in perinatology; 2000: Elsevier.

[28] Gandevani SB, Garshasbi A, Dibaj S. Cut-off value of 1-h, 50-g glucose challenge test for screening of gestational diabetes mellitus in an Iranian population. Journal of Obstetrics and Gynaecology Research. 2011;37(6):534-537.

[29] Sells CJ, Robinson NM, Brown Z, Knopp RH. Long-term developmental follow-up of infants of diabetic mothers. The Journal of pediatrics. 1994;125(1):S9-S17.

[30] Nold JL, Georgieff MK. Infants of diabetic mothers. Pediatric Clinics. 2004;51(3):619-637.

[31] Bánhidy F, Ács N, Puhó EH, Czeizel AE. Congenital abnormalities in the offspring of pregnant women with type 1, type 2 and gestational diabetes mellitus: A population-based case-control study. Congenital anomalies. 2010;50(2):115-121.

[32] Dafallah SE, Yousif EM. Diabetes mellitus during pregnancy. Fetal outcome. Saudi medical journal. 2004;25(12):2041-2042.

[33] Meur S, Mann NP. Infant outcomes following diabetic pregnancies. Paediatrics and Child Health. 2007;17(6):217-222.

[34] Kazandi M, Hasdemir P, Zeybek B, Akcay A. Placental growth factor: a putative screening test for gestational diabetes mellitus in first trimester. Clinical & Experimental Obstetrics & Gynecology. 2010;37(4):322.

[35] Eidelman AI, Samueloff A, editors. The pathophysiology of the fetus of the diabetic mother. Seminars in perinatology; 2002: Elsevier.

[36] Homko C, Sivan E, Chen X, Reece E, Boden G. Insulin secretion during and after pregnancy in patients with gestational diabetes mellitus. The

Journal of Clinical Endocrinology & Metabolism. 2001;86(2):568-573.

[37] Hami J, Shojae F, Vafaee-Nezhad S, Lotfi N, Kheradmand H, Haghir H. Some of the experimental and clinical aspects of the effects of the maternal diabetes on developing hippocampus. World journal of diabetes. 2015;6(3):412.

[38] Takata K, Hirano H. Mechanism of glucose transport across the human and rat placental barrier: a review. Microscopy research and technique. 1997;38(1-2):145-152.

[39] Shin B-C, Fujikura K, Suzuki T, Tanaka S, Takata K. Glucose transporter GLUT3 in the rat placental barrier: a possible machinery for the transplacental transfer of glucose. Endocrinology. 1997;138(9):3997-4004.

[40] Myers MG, White MF. The new elements of insulin signaling: insulin receptor substrate-1 and proteins with SH2 domains. Diabetes. 1993;42(5):643-650.

[41] Eriksson UJ. Diabetes in pregnancy: retarded fetal growth, congenital malformations and feto-maternal concentrations of zinc, copper and manganese in the rat. The Journal of nutrition. 1984;114(3):477-484.

[42] Moore D, Stanisstreet M, Clarke C, Beck F, Hipkin L. Effect on rat embryos of in vitro culture in sera from human diabetic patients. Diabetes research (Edinburgh, Scotland). 1987;5(3):139-43.

[43] Sadler T, Hunter E, Balkan W, Horton W. Effects of maternal diabetes on embryogenesis. American journal of perinatology. 1988;5(04):319-326.

[44] Horton W, Sadler T. Effects of maternal diabetes on early embryogenesis: alterations in morphogenesis produced by the ketone body, B-hydroxybutyrate. Diabetes. 1983;32(7):610-616.

[45] Lampl M, Jeanty P. Exposure to maternal diabetes is associated with altered fetal growth patterns: A hypothesis regarding metabolic allocation to growth under hyperglycemic-hypoxemic conditions. American Journal of Human Biology: The Official Journal of the Human Biology Association. 2004;16(3):237-263.

[46] Metzger B. The legacy of Norbert Freinkel: maternal metabolism and its impact on the offspring, from embryo to adult. Israel journal of medical sciences. 1991;27(8-9):425-431.

[47] Eriksson U, Borg L. Protection by free oxygen radical scavenging enzymes against glucose-induced embryonic malformations in vitro. Diabetologia. 1991;34(5):325-331.

[48] Kalkhoff RK. Impact of maternal fuels and nutritional state on fetal growth. Diabetes. 1991;40(Supplement 2):61-5.

[49] Afkhami M, Rashidi M. Gestational diabetes mellitus. Hormozgan Medical Journal. 2007;11(1):1-12.

[50] Eriksson U, Swenne I. Diabetes in pregnancy: growth of the fetal pancreatic B cells in the rat. Neonatology. 1982;42(5-6):239-248.

[51] Hay WW. Care of the infant of the diabetic mother. Current diabetes reports. 2012;12(1):4-15.

[52] Haworth J, McRae K, Dilling LA. Prognosis of infants of diabetic mothers in relation to neonatal hypoglycaemia. Developmental Medicine & Child Neurology. 1976;18(4):471-479.

[53] Mazzone D, Milana A, Grasso S, Nicotra C, Milana G, Dell'Aquila N. The newborn infant of the diabetic mother:

the clinical findings in 431 subjects. La Pediatria medica e chirurgica: Medical and surgical pediatrics. 1993;15(3):257-261.

[54] Correa A, Gilboa S, Besser L, Botto L, Moore C, Hobbs C, et al. Diabetes mellitus and birth defects. Obstetric Anesthesia Digest. 2009;29(1):40-41.

[55] Konarzewska J, Wójcikowski C. Risk of diabetes mellitus after pregnancy complicated by gestational diabetes mellitus (GDM). Ginekologia polska. 2004;75(10):754-759.

[56] Eriksson RS, Thunberg L, Eriksson UJ. Effects of interrupted insulin treatment on fetal outcome of pregnant diabetic rats. Diabetes. 1989;38(6):764-772.

[57] Farrell T, Neale L, Cundy T. Congenital anomalies in the offspring of women with type 1, type 2 and gestational diabetes. Diabetic Medicine. 2002;19(4):322-326.

[58] Hami J, Hosseini M, Ivar G, Vafaei-Nezhad S, Keivan M. Cognitive Function in Offspring of Mothers with Gestational Diabetes–The Role of Insulin receptor. MOJ Anat & Physiol. 2016;2(7):00072.

[59] Fuhrmann K, Reiher H, Semmler K, Fischer F, Fischer M, Glöckner E. Prevention of congenital malformations in infants of insulin-dependent diabetic mothers. Diabetes care. 1983;6(3):219-223.

[60] Suhonen L, Hiilesmaa V, Teramo K. Glycaemic control during early pregnancy and fetal malformations in women with type I diabetes mellitus. Diabetologia. 2000;43(1):79-82.

[61] Mohammad-Beigi A, Tabatabaee SHR, Yazdani M, Mohammad-salehi N. Gestational diabetes related unpleasant outcomes of pregnancy. KAUMS Journal (FEYZ). 2007;11(1):33-38.

[62] Yamashita Y, Kawano Y, Kuriya N, Murakami Y, Matsuishi T, Yoshimatsu K, et al. Intellectual development of offspring of diabetic mothers. Acta Paediatrica. 1996;85(10):1192-1196.

[63] Rizzo T, Metzger BE, Burns WJ, Burns K. Correlations between antepartum maternal metabolism and intelligence of offspring. New England Journal of Medicine. 1991; 325(13):911-916.

[64] Delascio Lopes C, Sinigaglia-Coimbra R, Mazzola J, Camano L, Mattar R. Neurofunctional evaluation of young male offspring of rat dams with diabetes induced by streptozotocin. International Scholarly Research Notices. 2011;2011.

[65] Ornoy A, Ratzon N, Greenbaum C, Peretz E, Soriano D, Dulitzky M. Neurobehaviour of school age children born to diabetic mothers. Archives of Disease in Childhood-Fetal and Neonatal Edition. 1998;79(2):F94-FF9.

[66] Vafaei-Nezhad S, Hami J, Sadeghi A, Ghaemi K, Hosseini M, Abedini M, et al. The impacts of diabetes in pregnancy on hippocampal synaptogenesis in rat neonates. Neuroscience. 2016;318:122-133.

[67] Carrapato M, Marcelino F. The infant of the diabetic mother: The critical developmental windows. Early pregnancy (Online). 2001;5(1):57-58.

[68] Georgieff MK. The effect of maternal diabetes during pregnancy on the neurodevelopment of offspring. Minnesota medicine. 2006;89(3):44-47.

[69] Akyol A, Kiylioglu N, Bolukbasi O, Guney E, Yurekli Y. Repeated hypoglycemia and cognitive decline. A case report Neuroendocrinol Lett. 2003;24:54-56.

[70] Smoak IW, Sadler TW. Embryopathic effects of short-term exposure to hypoglycemia in mouse embryos in vitro. American journal of obstetrics and gynecology. 1990;163(2):619-624.

[71] Rizzo T, Silverman B, Metzger B, Cho N. Behavioral adjustment in children of diabetic mothers. Acta paediatrica. 1997;86(9):969-974.

[72] Babiker OO, Stathupulo E. Long-term effects of maternal diabetes on their offsprings development and behaviours. Sudanese Journal of Paediatrics. 2007;8(1):133-146.

[73] Ornoy A, Ratzon N, Greenbaum C, Wolf A, Dulitzky M. School-age children born to diabetic mothers and to mothers with gestational diabetes exhibit a high rate of inattention and fine and gross motor impairment. Journal of Pediatric Endocrinology and Metabolism. 2001;14(Supplement):681-90.

[74] Hami J, Vafaei-Nezhad S, Ivar G, Sadeghi A, Ghaemi K, Mostafavizadeh M, et al. Altered expression and localization of synaptophysin in developing cerebellar cortex of neonatal rats due to maternal diabetes mellitus. Metabolic brain disease. 2016;31(6):1369-1380.

[75] Nelson CA, Wewerka S, Thomas KM, deRegnier R-a, Tribbey-Walbridge S, Georgieff M. Neurocognitive sequelae of infants of diabetic mothers. Behavioral neuroscience. 2000;114(5):950.

[76] Rizzo TA, Metzger BE, Dooley SL, Cho NH. Early malnutrition and child neurobehavioral development: insights from the study of children of diabetic mothers. Child development. 1997;68(1):26-38.

[77] Lagace DC, Donovan MH, DeCarolis NA, Farnbauch LA, Malhotra S, Berton O, et al. Adult hippocampal neurogenesis is functionally important for stress-induced social avoidance. Proceedings of the National Academy of Sciences. 2010;107(9):4436-4441.

[78] Kinney B, Rabe M, Jensen R, Steger R. Maternal hyperglycemia leads to gender-dependent deficits in learning and memory in offspring. Experimental Biology and Medicine. 2003;228(2):152-159.

[79] DeBoer T, Wewerka S, Bauer PJ, Georgieff MK, Nelson CA. Explicit memory performance in infants of diabetic mothers at 1 year of age. Developmental medicine and child neurology. 2005;47(8):525-531.

[80] Sadeghi A, Asghari H, Hami J, Roodi MM, Mostafaee H, Karimipour M, et al. Volumetric investigation of the hippocampus in rat offspring due to diabetes in pregnancy–A stereological study. Journal of chemical neuroanatomy. 2019;101:101669.

[81] Tehranipour M, Khakzad M. Effect of maternal diabetes on hippocampus neuronal density in neonatal rats. J Biol Sci. 2008;6:1027-1032.

[82] Garner CC, Zhai RG, Gundelfinger ED, Ziv NE. Molecular mechanisms of CNS synaptogenesis. Trends in neurosciences. 2002;25(5):243-250.

[83] Südhof TC. The synaptic vesicle cycle. Annu Rev Neurosci. 2004;27:509-547.

[84] Chambers JS, Thomas D, Saland L, Neve RL, Perrone-Bizzozero NI. Growth-associated protein 43 (GAP-43) and synaptophysin alterations in the dentate gyrus of patients with schizophrenia. Progress in Neuro-Psychopharmacology and Biological Psychiatry. 2005;29(2):283-290.

[85] Frick K, Fernandez S, Bulinski S. Estrogen replacement improves spatial reference memory and increases hippocampal synaptophysin in aged female mice. Neuroscience. 2002;115(2):547-558.

[86] Frick KM, Fernandez SM. Enrichment enhances spatial memory and increases synaptophysin levels in aged female mice. Neurobiology of aging. 2003;24(4):615-626.

[87] Sze C-I, Troncoso JC, Kawas C, Mouton P, Price DL, Martin LJ. Loss of the presynaptic vesicle protein synaptophysin in hippocampus correlates with cognitive decline in Alzheimer disease. Journal of Neuropathology & Experimental Neurology. 1997;56(8):933-944.

[88] Sadeghi A, Esfandiary E, Hami J, Khanahmad H, Hejazi Z, Mardani M, et al. The effects of maternal diabetes and insulin treatment on neurogenesis in the developing hippocampus of male rats. Journal of chemical neuroanatomy. 2018;91:27-34.

[89] Sadeghi A, Esfandiary E, Hami J, Khanahmad H, Hejazi Z, Razavi S. Effect of maternal diabetes on gliogensis in neonatal rat hippocampus. Advanced biomedical research. 2016;5.

[90] Lotfi N, Hami J, Hosseini M, Haghir D, Haghir H. Diabetes during pregnancy enhanced neuronal death in the hippocampus of rat offspring. International Journal of Developmental Neuroscience. 2016;51:28-35.

[91] Haghir H, Hami J, Lotfi N, Peyvandi M, Ghasemi S, Hosseini M. Expression of apoptosis-regulatory genes in the hippocampus of rat neonates born to mothers with diabetes. Metabolic brain disease. 2017;32(2):617-628.

[92] Sardar R, Hami J, Soleimani M, Joghataei M-T, Shirazi R, Golab F, et al. Maternal diabetes-induced alterations in the expression of brain-derived neurotrophic factor in the developing rat hippocampus. Journal of Chemical Neuroanatomy. 2021;114:101946.

[93] Kamal A, Biessels G-J, Gispen WH, Ramakers GM. Synaptic transmission changes in the pyramidal cells of the hippocampus in streptozotocin-induced diabetes mellitus in rats. Brain research. 2006;1073:276-280.

[94] Revsin Y, Saravia F, Roig P, Lima A, de Kloet ER, Homo-Delarche F, et al. Neuronal and astroglial alterations in the hippocampus of a mouse model for type 1 diabetes. Brain research. 2005;1038(1):22-31.

[95] Copp AJ, Greene ND. Genetics and development of neural tube defects. The Journal of Pathology: A Journal of the Pathological Society of Great Britain and Ireland. 2010;220(2):217-230.

[96] Stranahan AM, Arumugam TV, Cutler RG, Lee K, Egan JM, Mattson MP. Diabetes impairs hippocampal function through glucocorticoid-mediated effects on new and mature neurons. Nature neuroscience. 2008;11(3):309-317.

[97] Kamal A, Biessels G-J, Urban I, Gispen W. Hippocampal synaptic plasticity in streptozotocin-diabetic rats: impairment of long-term potentiation and facilitation of long-term depression. Neuroscience. 1999;90(3):737-745.

[98] Jackson-Guilford J, Leander JD, Nisenbaum LK. The effect of streptozotocin-induced diabetes on cell proliferation in the rat dentate gyrus. Neuroscience letters. 2000;293(2):91-94.

[99] Ho N, Sommers MS, Lucki I. Effects of diabetes on hippocampal neurogenesis: links to cognition and depression. Neuroscience & Biobehavioral Reviews. 2013;37(8):1346-1362.

[100] Altunkaynak B, Unal D, Altunkaynak M, Halici Z, Kalkan Y, Keles O, et al. Effects of diabetes and ovariectomy on rat hippocampus (a biochemical and stereological study). Gynecological Endocrinology. 2012;28(3):228-233.

[101] Liu T, Bai Y, Ma L, Ma X, Wei W, Zhang J, et al. Altered Effective Connectivity of Bilateral Hippocampus in Type 2 Diabetes Mellitus. Frontiers in neuroscience. 2020;14:657.

[102] Saravia FE, Revsin Y, Deniselle MCG, Gonzalez SL, Roig P, Lima A, et al. Increased astrocyte reactivity in the hippocampus of murine models of type 1 diabetes: the nonobese diabetic (NOD) and streptozotocin-treated mice. Brain research. 2002;957(2):345-353.

[103] Lebed YV, Orlovsky MA, Nikonenko AG, Ushakova GA, Skibo GG. Early reaction of astroglial cells in rat hippocampus to streptozotocin-induced diabetes. Neuroscience letters. 2008;444(2):181-185.

[104] Pamidi N, BN SN. Effect of streptozotocin induced diabetes on rat hippocampus. Bratislavske lekarske listy. 2012;113(10):583-588.

[105] Baydas G, Nedzvetskii VS, Nerush PA, Kirichenko SV, Yoldas T. Altered expression of NCAM in hippocampus and cortex may underlie memory and learning deficits in rats with streptozotocin-induced diabetes mellitus. Life sciences. 2003;73(15):1907-1916.

[106] Baptista FI, Pinto MJ, Elvas F, Almeida RD, Ambrósio AF. Diabetes alters KIF1A and KIF5B motor proteins in the hippocampus. PloS one. 2013;8(6):e65515.

[107] Piotrowski P, Wierzbicka K, Smialek M. Neuronal death in the rat hippocampus in experimental diabetes and cerebral ischaemia treated with antioxidants. Folia neuropathologica. 2001;39(3):147-154.

[108] He X, Sun J, Huang X. Expression of caspase-3, Bax and Bcl-2 in hippocampus of rats with diabetes and subarachnoid hemorrhage. Experimental and therapeutic medicine. 2018;15(1):873-877.

[109] Copp AJ, Brook FA, Estibeiro JP, Shum AS, Cockroft DL. The embryonic development of mammalian neural tube defects. Progress in neurobiology. 1990;35(5):363-403.

[110] Greene ND, Copp AJ. Development of the vertebrate central nervous system: formation of the neural tube. Prenatal Diagnosis: Published in Affiliation with the International Society for Prenatal Diagnosis. 2009;29(4):303-311.

[111] Parchem RJ, Moore N, Fish JL, Parchem JG, Braga TT, Shenoy A, et al. miR-302 is required for timing of neural differentiation, neural tube closure, and embryonic viability. Cell reports. 2015;12(5):760-773.

[112] Dessaud E, McMahon AP, Briscoe J. Pattern formation in the vertebrate neural tube: a sonic hedgehog morphogen-regulated transcriptional network. 2008.

[113] Moore LL, Singer MR, Bradlee ML, Rothman KJ, Milunsky A. A prospective study of the risk of congenital defects associated with maternal obesity and diabetes mellitus. Epidemiology. 2000:689-694.

[114] Salbaum JM, Kappen C. Neural tube defect genes and maternal diabetes during pregnancy. Birth Defects Research Part A: Clinical and Molecular Teratology. 2010;88(8):601-611.

[115] Meier P, Finch A, Evan G. Apoptosis in development. Nature. 2000;407(6805):796-801.

[116] Avagliano L, Doi P, Tosi D, Scagliotti V, Gualtieri A, Gaston-Massuet C, et al. Cell death and cell proliferation in human spina bifida. Birth Defects Research Part A: Clinical and Molecular Teratology. 2016;106(2):104-113.

[117] Mandelbrot L, Legardeur H, Girard G. Screening for gestational diabetes mellitus: Is it time to revise the recommendations? Gynecologie, obstetrique & fertilite. 2010;38(6):409-414.

[118] Hiéronimus S, Le Meaux J. Relevance of gestational diabetes mellitus screening and comparison of selective with universal strategies. Diabetes & metabolism. 2010;36(6 Pt 2):575-586.

[119] Khaksar Z, Jelodar G, Hematian H. Cerebrum malformation in offspring of diabetic mothers. Comparative Clinical Pathology. 2012;21(5):699-703.

[120] Hami J, Vafaei-Nezhad S, Ghaemi K, Sadeghi A, Ivar G, Shojae F, et al. Stereological study of the effects of maternal diabetes on cerebellar cortex development in rat. Metabolic brain disease. 2016;31(3):643-652.

[121] Coughlan M, Oliva K, Georgiou H, Permezel J, Rice G. Glucose-induced release of tumour necrosis factor-alpha from human placental and adipose tissues in gestational diabetes mellitus. Diabetic Medicine. 2001;18(11):921-927.

[122] Georgieff MK. Long-term brain and behavioral consequences of early iron deficiency. Nutrition reviews. 2011;69(suppl_1):S43-S8.

[123] Armony-Sivan R, Eidelman AI, Lanir A, Sredni D, Yehuda S. Iron status and neurobehavioral development of premature infants. Journal of perinatology. 2004;24(12):757-762.

[124] Plagemann A, Harder T, Lindner R, Melchior K, Rake A, Rittel F, et al. Alterations of hypothalamic catecholamines in the newborn offspring of gestational diabetic mother rats. Developmental Brain Research. 1998;109(2):201-209.

[125] Xiang AH, Wang X, Martinez MP, Walthall JC, Curry ES, Page K, et al. Association of maternal diabetes with autism in offspring. Jama. 2015;313(14):1425-1434.

[126] Hami J, Vafaei-Nezhad S, Haghir D, Haghir H. Insulin-like growth factor-1 receptor is differentially distributed in developing cerebellar cortex of rats born to diabetic mothers. Journal of Molecular Neuroscience. 2016;58(2):221-232.

[127] Haghir H, Sankian M, Kheradmand H, Hami J. The effects of induced type-I diabetes on developmental regulation of insulin & insulin like growth factor-1 (IGF-1) receptors in the cerebellum of rat neonates. Metabolic brain disease. 2013;28(3):397-410.

[128] Joca SR, Guimarães FS, Del-Bel E. Inhibition of nitric oxide synthase increases synaptophysin mRNA expression in the hippocampal formation of rats. Neuroscience letters. 2007;421(1):72-76.

[129] Gincel D, Shoshan-Barmatz V. The synaptic vesicle protein synaptophysin: purification and characterization of its channel activity. Biophysical journal. 2002;83(6):3223-3229.

[130] Zhan S-S, Beyreuther K, Schmitt H. Quantitative Assessment of the Synaptophysin Immuno-Reactivity of the Cortical Neuropil in Various Neurodegenerative Disorderswith Dementia. Dementia and Geriatric Cognitive Disorders. 1993;4(2):66-74.

[131] Eastwood S, Burnet P, Harrison P. Altered synaptophysin expression as a marker of synaptic pathology in schizophrenia. Neuroscience. 1995;66(2):309-319.

[132] Vawter M, Thatcher L, Usen N, Hyde T, Kleinman J, Freed W. Reduction of synapsin in the hippocampus of patients with bipolar disorder and schizophrenia. Molecular psychiatry. 2002;7(6):571-578.